MILITANCY AND CONFLICT RESOLUTION IN MANIPUR

MILITANCY AND CONFLICT RESOLUTION IN MANIPUR

Comdt. R S Jassal (Retd)

AKANSHA PUBLISHING HOUSE
NEW DELHI • GUWAHATI • VISAKHAPATNAM

AKANSHA PUBLISHING HOUSE
4649-B/21, Ansari Road
Darya Ganj, New Delhi - 110 002
Ph.: 23263193 / 9811582579 / 09435116718
Fax: 011-23263193
Email: ektabooks@yahoo.com
www.akanshapublishinghouse.com

Militancy and Conflict Resolution in Manipur

First Published 2011
ISBN 978-81-8370-285-0

PRINTED IN INDIA

Published by Akansha Publishing House, New Delhi and Printed at Harmain Offset Press, Delhi.

Dedicated

to

My mother late Sardarni Amar Kaur & my grand father late Lala Basant Ram whose dedicated efforts made me obtain Master's Degree in Sanskrit in absence of my father late Sardar Bhagat Singh who left for his heavenly abode at much early age when he was just 28 yeas old.

Contents

Foreword

'Militancy and Conflict Resolution in Manipur' is an anthology of the Thoughts of Commandant R.S. Jassal (Retd) who is a soldier turned into a writer and social activist. I have seen the transformation of Commandant Jassal from an officer of the Indian Army to a responsible senior citizen of the country.

Commandant Jassal's excellent work is a collection of very learned papers of the author written on different occasion, his immediate, not so immediate and long term analysis of the issues that have confronted this part of the country, especially the rugged hill country of Nagaland and a complex social milieu known as Manipur.

The core contents of the book consist of seventeen essays dealing with the issues of conflict in Manipur and Nagaland. Commandant Jassal writes as an expert, a man who knows the two states from the grass roots. There is as provocative dialogue of Commandant Jassal with a young Naga research scholar on the sensitivity of the Meitei - Naga relationship. This dialogue draws the best out of Commandant Jassal. He writes on the Naga attitudes; The Naga Tangle; Naga Intransigence. The Myanmarese and Non-Myanmarese factors affecting Manipur are of good interest. He boldly analyses the cultural polarization in Manipur which is a reflection of the pluralism and modernization in the state. He also touches on the staggering Human Right issues deftly.

The quality of the book is increased by two tables. First, namely Intrinsic Conflict: State-wise List of Outfits with Last Known Strength and Number of Weapons. Second, Extrinsic Conflict: Myanmarese Outfits, Existing Under Ceasefire Agreement and Those Operating Outside any Ceasefire Agreement. The information that emanate from the two tables will be useful for the students of the problem and policy analysts and planners.

The references and notes contain eleven hot problems, ranging from the Independence Movements: Naga Hills to The Ghosts of History, Porous Borders in Manipur insurgency, Intractable Loktak as Achille's Heels and Redefining Conflict.

A very learned Glossary was provided by the author which shows his intellectual depth. This book contains six appendices on issues like Conflict Mapping, The proclamation of Maharaja Bodhchandra Singh on October, 18, 1948 and the oft discussed "Merger Agreement".

Reading into these essays and notes, one feels that the author is a knowledgeable writer with full knowledge of history, culture and political development of north-eastern frontier, particularly Manipur and Nagaland. Anybody, both the common man and the scholarly will enjoy reading this work of Commandant Jassal. I congratulate him for taking the trouble of putting all his learned and thought-provoking writings into a book. I wish him success in his life.

Gangmumei Kamei

National Fellow,

Indian Institute of Advanced Study

Shimla.

Preface

"In spite of great funds of knowledge in all faiths, men have been unable to reconcile their differences causing great conflicts, wars and suffering" Today ***'I am',*** tomorrow ***'you will be'***. What happens in between the metaphorical ***'today'*** and ***'tomorrow'*** -are multi dimensional events which impact the ethnicities, societies and nations as a whole. Some lapsing into the graveyards/cemeteries unrevealed if untold, some fading with the memories after oral releases carried through oral histories and some recorded in black and white which assume the status of histories. Some histories go into revisionist status due certain political, industrial, economic revolutions purportedly dictated at the whims and fancies of the rulers of the time to replace some hard facts. This revisionist history gets offender's status because it is manipulated to suit the agenda of a few.

This is exactly what was in the most of the histories of NE States and Manipur in particular. ***Puyas*** in Manipur were the only authoritative events of history which were also found by then Maharaja Garibniwaz as misleading and lacking authentic details so he ordered their burning during 1724. British administrators recorded facts as true as possible in Government Gazettes; history books and personal memoirs which became the naturally dependable blocks to build up the sustainable knowledge edifices which affected destinies of people divided in Village states and Salai states.

By virtue of my 34 years of active service with the Assam Rifles, repeated postings into Nagaland, Manipur, Arunachal Pradesh and

Mizoram, intrinsic internal inquiry as to what went wrong and where, why dissatisfaction with our constitutional systems of India, why so many groups fighting contradictory battles, did political leadership of Centre (1947) not realize at that time what should have been the genuine lines of demarcation of two nations and for carving new States and so many such and also unrelated queries, took me to the people's thresholds into their private chambers of oral histories, and living memories of unforgettable star personalities to mention a few like late Chuba Sangtam of Tuensang proper who had claimed himself having killed 18 Phizo-ites on their entry into Tuensang area, on creation of NHTA (Naga Hills & Tuensang Area) he was appointed as first Base Supdt (illiterate though) being a towering personality; and late Thungdti Chang the Ahng of Hongkin State (consisting of Tuensang, Mon and Kiphire present day districts of Nagaland) and also late Baba Chuthang Konyak of Monyakshu (now Mon District) that I could get some revealing facts which factored into writing of this book. I have travelled tirelessly into interiors of Manipur for past 14 years since retirement (1997) yet my journeys remain incomplete to grasp the vastness of this tiny state as topography and people are just wonderful and too great!

A word about my personal space into Manipuri society. My marriage with a Tangkhul girl whose father late C. Athuk, a school teacher was warden of both late Shri Y. Shaiza and Shri Rishang Keishing, legendary duo - both Tangkhul who never met each other during political career, two daughters from us married into Meitei families, — one in Sougaijam clan (a complete convert and most respectable clan from Kuki to Meitei) and the other from Mairembam clan of Moirang which produced a Congress INA turned politician most proud Indian, a shrewd popular leader late M. Koireng Singh as the CM, a clan from a salai who though related to Rajas/Maharajas of Imphal but always remained on fight modes till both merged into one as Meitei and established Kangla as their final Kingdom. Son got married into Rangteinamei clan of Maram tribe from Maram Khullen whose father-in-law Mr. Nupni Bata was killed by NA (Naga Army) UGs (Undergrounds) NFG (Naga Federal Government) in 1968 and mother-in-law late D'Kula Mary *(a manly woman)* first Maram graduate who established D'Regina English High school at Chingmeirong Imphal. With the death of my wife I got remarried in another very good and noble Kuki family from Kipgen clan at T/Waichong (Sadar Hills). All these family ties gave me an edge over other researchers to get the root based replies to my queries from needle to elephant or say onto anything, I wished to know.

Then my personal links with various public leaders and political personalities from Nagaland, like Shri SC Jamir former Chief Minister Nagaland and former Governor of Goa-Maharashtra, Manipur RK Dorendro Singh and NC Hills late Lenjao Hmar noted scholars and historians and long stint in the Assam Rifles gave me deep understanding of the people's problems, irks and at times misconceptions too. So whatever information and data I could store from various sources I wanted to collate and put up in one place before going it as a 'lapsed knowledge', hence this humble attempt. In the remedial measures, I have realized it is only police, bureaucrats and sincere politicians who can bring a change. There can be no better social worker than this trio. They have vision, strength, resources and stakes too all put in one, on their side. Whatever they wish to give to the society will come back to them ultimately. Military is only a stop gap filler to bring derailed administrative controls back to rails, to assist administration to do their act though nevertheless, Military role has been and will remain extremely honoured in the Nation Building as seen from their role played in 1947 partition and 1971 Bangla Operations. Currently role in the NE is again something superb and highly commendable. It is only fair and dare governance which is needed by the people. They demand nothing more. It was cyclic development under British rulers who tried to bring fragmented but all very assertive, independent and powerful groups and tribes to a disciplined and collective set of governance under one administrative head i.e., DC (Deputy Commissioner). Manipur was a single DC administrated state till as late as 1970. It was a slow and steady development from one district to nine districts. Independence was of entirely a different reading in this region which gave opportunity to administrators- politicians to continue with whatever vision they had but less one which stood very loud and clear i.e., the least SERVICE TO THE PEOPLE. Constitutional guarantees failed to catch up with the changing times due to lack of education, communications and awareness amongst tribal's, hence re-emergence of ETHNICITY STRENGTHS ONLY started to be considered as an alternative force for garnishing political strength which meant economic strength for a few countables. Primordial religious concepts badly messed into Vaishnaivism and Christian beliefs which contributed though to enlightenment and material enhancement but gradually and unconscientiously transformed once homogenous societies into heterogeneous ones which we confront today.

With this background in mind, I have put in sincere, undiluted and bold attempt to chronicle the basic reasons of dissent and discontent leading to current position of Nagas feeling safe only if all get under one administrative umbrella though in reality a myth; Kuki's Hmars-Zous also same way for ZO-Ram (Mizoram inclusive) Brus running homeless and Meitei's taking everything only with integrity of Manipur territories as graphed by the British not to be lost even to the last ounce of blood of each Meitei. This problem can be resolved by new generation players if they keenly study the happenings between ***'today's'*** and ***'tomorrow'*** as pointed out in the beginning of this preface with sincerity and focused vision. I do appreciate to free our people from the present ***'bondage'*** and meet their aspirations, the youngsters of today from various quarters have initiated series of initiatives and workshops with serious interest, My objective is to harness what Tertullian said: ***"The entire fruit is already in the seed"***. I want to nurture this signal of hope carefully through mass provocation of fertile minds. Now is NOT the time to find each other's faults but to find a solution.

RS JASSAL

Chingmeirong West
Behind Police outpost
Imphal, Manipur - 795001

Acknowledgements

This book would remain incomplete, if I donot thank Prof. Gangmumei Kamei, for encouraging me to bring various articles and my studies into book form; Prof. Sharma for his kind editing; my son, Harkiran Jassal who has always to bear with my demand over demand to type drafts. Special thanks to Mariembam Mani Singh my noble brother-in-law for constant moral support and to Akansha Publishing House to give brilliant shape and style to this book and for wonderful thought for Shri SC Jamir, Lt. Gen. NK Singh UYSM, AVSM, USM and Prof. Laldena.

लेफ्टिनेंट जनरल एन के सिंह, यू वाई एस एम, वी एस एम
जनरल अफसर कमाडिंग
Lt Gen NK Singh, UYSM, AVSM, VSM
General Officer Commanding

मुख्यालय ३ कोर
पिन - ९०८५०३
द्वारा ९९ सेना डाकघर
HQ 3 Corps
PIN - 908503
C/o 99 APO

A Word About Mr Jassal

Mr. R.S. Jassal devoted his lifetime for the interest of the people for bringing peace and happiness in the State of Manipur. Despite being more than 70 yrs old, he has, untiringly continued his efforts through various means like print and electronic media, seminars, discussions and public forums to maintain tranquility in Manipur and bring about development. His enthusiastic interest has gone a long way in rehabilitating the disabled, deprived and poorer sections of the society in Manipur. His great efforts have contributed very meaningfully in achieving the overall improvements in Manipur in various fields over the past few years.

I wish him good luck and success in all his future endeavours.

14 Feb 2011

(NK Singh)
Lt Gen
GOC 3 Corps

As I Know Col. (retired) Jassal

In the sixties I was elected as the first Lok Sabha Member from Nagaland. In those days only a handful of Nagas were to be found at New Delhi. Even as a Member of Parliament, I felt as if I was staying in a foreign land. At that time by coincidence my wife saw a rare Naga face in INA market at Delhi and she happened to be the first late wife of Capt. Jassal. She was brought to my residence at 12 Tuglak Road by my wife. Since then I know Jassal through this wonderful lady.

After he had joined Assam Rifles and posted in Nagaland, I had been watching his performance from very close quarter. His service rendered to the forces he had joined, I am sure had been immense because of his intimate knowledge of the area. He is the one who had fully identified with the local inhabitants not because he married local girl but because of his endeavour to assimilate with the local environment and traits. He is a determined person who had assiduously worked as a journalist and a writer. His knowledge about the Naga problem fortified by his wisdom deserves careful consideration by all who are engaged in finding solution to this vex problem. I am sure that he will continue to render his invaluable service to the people of this region and the country.

S.C. Jamir

Former Chief Minister of Nagaland
and
Former Governor of Goa and Maharashtra

25-2-2011

27/2/2011

Rattan Singh Jassal, Commdt (Retd): The Unique Case of Tribalization of a Sardarji

With brilliant academic career securing Gold Medal in Sanskrit in his post-graduation, and many medals for his meritorious services. R.S. Jassal retired in the rank of Commandant Assam Rifles, MHA. In spite of such brilliant academic career and colorful military service records, Jassalji has no air at all and is so modest and warm-hearted. In short, he is a man full of milk of human kindness. I had a privilege of associating myself with him during my seminar and workshop on Tipaimukh dam at Churachandpur and Ruonglevaisuo (Tipaimukh). We were given warm receptions in all military camps from Churachandpur town to Tipaimukh. This shows how much respect and love Jassalji has still been enjoying from his former colleagues. One of the rarest qualities of Jassalji is his ardent love of the people and his wide knowledge of the multi-faced problems of the region. Looking at his family background, Jassalji is a replica of national integration - his first (late) wife being a Tangkhul lady and the present one a Kipgen of the Thadou Kuki. Both his two daughters married Meitei and his son a Maram Naga lady. Jassalji is therefore a unique case of tribalization of a Sardarji.

11/3/2011

Prof. Lal Dena
Manipur University

1

ASYMMETRIC CHANGE

Ancient polities used to be simplistic and without complexities–the law of unintended consequences still not enunciated–fundamentally because almost everybody would be in a state of *hrideya* (or affectionate towards the 'other'). As between 'one' and the other' there used to be no conflict of interest so they would not lock horns. One has seen such Gestalt psychology in King Lear, surrendering to the 'other' or Othello dying locked up on his own. Unless there is a holistic stance of organically looking at the total, one would never feel rationalistic enough to integrate in the 'whole.' In fact, *Samâna* = *(sa + mânas)* is of the nature being *hrideya;* hence samâna + hrideya = *sahrideya* is an active present participle formed from the root of 'to be'; so that it means 'being for oneself; and this one: 'self' is in harmony (or symmetry) with the 'other self'. Luckily put, this self (or one) often remains in sync or symmetrical with the other at least in recent past, when networking among people used to be rather simple.

In fact, something else happens whenever the two falls out of sync. Paradoxically even the empirical author 'I' sink into the 'other' [reader-spectator-audience] a new something *(rasa)* comes into play between the author (actor) and the reader (spectator). And when these two separate beings could be brought into sync, or onto symmetry or harmony, the. two are called *samâna* + *hrdeya* = *sahrideya.* One is the performer; spectator (other) is the *hrideya. Rasa,* the *dhvani* complex, mediating language and the psychic communion (+whatever harmonic; values obtained as well as body language, chemistry) are what transpires from 'one' (dramatist) to the 'other' (audience). The enjoyer of *rasa* is called the *'rasika'*. Alternatively put, *'shrestha'* whenever the 'particular' can evoke the 'general' or the 'other'. Following Gestalt psyche, such harmony or symmetry renders an extra 'being' over and above the sum of all the parts, making the whole.

There may be occasions when the 'other' feels 'unsettled' while 'one' shows intent and ambition. Interestingly up to that tipping point, some kind of 'hideous' equity prevails (if not symmetry).[1]

> "If you cross the North Korean border illegally, you get 12 years hard labour. If you cross the Iranian border illegally, you are detained indefinitely. If you cross the Afghan border illegally, you get shot. If you cross the Saudi Arabian border, you will be jailed. If you cross the Chinese border illegally, you will never be heard from again. If you cross the Venezuelan border illegally you will be branded a spy and your fate will be sealed .... If you cross the Indian border, you get a ration card, voter identification, passport, Haj subsidy, a job, a driver's licence, an identity card, job reservation, special privileges, credit cards, subsidised rent or a loan to buy a house, free education, free health, a lobbyist in New Delhi and voting rights. Welcome to India."
> There are yet other occasions when the emotions overflow from the spectators in an super-surcharged setting as a cricket match, a one-day international (ODI) at that between India and South Africa (SA). Critically poised, India [the 'other'] makes a "classic" effort (February 14, 2010) at Eden Gardens, Kolkata to outreach SA's 2 wickets for 233 runs. As if the so-called 'punch line' was delivered by a sophisticated elderly aunt to an off-colour joke with obvious riles and gusto."

The history of any nation–people or race follows a chequered pattern of existentialism. There are periods of stagnations, bitter confrontations as well as supreme glory depending upon the path, channelised, dictated and followed by that particular race. History is such a natural flow of events, it does not halt. "Change is the continuous process like the course of a river. There is neither end of change nor end of history."[2] It keeps on occurring every moment here, there, everywhere. It is for the historians, writers, analysts and the whole elite to interpret based on the actual events as they view. These are periods of supremacy as well as of compulsive (ness) of stability and turmoil actuated by some, but attributed to 'others' (to make up a good majority) and thus made attributable (or acceptable) to all. Previewed situation prevalent in Manipur today have put its people in a complex and entwined bind of all sorts: factual, imageries, real + unreal, suspicions, superimpositions (which cannot be refuted or denied by the victim-target. The resulting scenario is sheer helplessness, imbued despondency, pushback, mal-development, social insecurity, emotional and financial breakdown, compounded by the presence of insurgent outfits, proliferating on an industrial scale with no end in sight.

2

CONFLICT *PER SE*

Human history has seen a paradigmatic shift in the concept, ormat, content, parameters, albeit purpose of 'conflict.' In those very ancient days one could imagine more of hand-to-hand fights among two: the hunter-predator claiming his prey in a 'hot chase' onto other's domain and the 'other' counterclaiming his find. But over the years it has morphed onto the most violent '*jihadi* attacks,' causing hitherto-unimagined maximum destruction at the spot-on-target, with the *jihadi* himself guiding and in the process getting himself exploded into smithereens. Yet two co-axial factors transcended: 'self' and, of course, the 'other.' Seen all through is the innate human process of 'othering', with plausible sample arguments like: 'My' prey is most certainly 'mine', and any 'other' can have it over 'my' dead body!" which is retorted by: "'You' might shout, but 'I' have had it in 'my' domain!" One way of looking at it surely spells out an irresistible: "Might is right!" or "Surely 'you' will have to fight to dispossess 'me'." and for either "It's not a 'win-win' situation!" The first dissent from symmetry is: 'I am not you! neither you me!'; once a discord breaks out, full scale confrontation may soon follow. Almost the same gamesmanship ('win-lose' trait) is crystal-clear in a terror-strike which now redefines 'conflict' as are all others external or instrumental to mankind during his myriad existence either for better or for worse. If earlier conflicts used to be more open, for instance actual fighting would be limited to the daytime, then waiting for another daybreak, it has since started becoming much more-stealth-based, as a matter of strategy, by using costly stealth-frigates for navy or stealth-bombers for air force, including surprise-attacks, IEDs/kuman bomb and unspecified . . . for greater devastation and 'sure win.' In plain terms, the humane milk was conspicuous by its presence at least in case of invaders like king Alexander the Great, who showed a still proverbial (near-mythical) mercy to his subdued enemies like king Porus amidst his equally ruthless campaign (of mindless killing). Yet its latest version of terror is devoid of any such humanity; in lieu of minimum devastation as of then, now the shift is to the other extreme: maximum devastation.[3]

In sharp contrast, its modern replica is: that a fundamentalist organization recruits *jihadis* or suicide-attackers (or volunteer-pilots to crash hijacked passenger aeroplanes for no other conceivable reason but to bring about a catastrophe) input-output ratio of: 2 against 2,000 lives, killed as in that US Twin-Tower: World Trade Center attack in New York in 2001. Notwithstanding its mammoth impact, it created a traumatic stress to the entire humanity (except perhaps these very hard core fundamentalists). Without doubt, the *jihadi* got a price from the terrorist organization ahead of his mission of the mass killing, resultant trauma, and revenge–a sure 'win-win' game. As a matter of sheer coincidence, this unprecedented event has perhaps unintentionally helped in triggering off an economic boom. For, soon afterwards, US experienced a sub-prime housing credit-boom which triggered a global boom, although 7 years later followed by a glut (2008) initially in the US and later enveloping the rest of the globe–the worst since the Great Depression of the 1930s. By no means, terror in any part of this digitally-connected, globalized world–where the largest (market economy) of the US is, as it were the engine of the global train to which are "coupled" other nations like China and India like railway bogies–was seen as a "game-changer" in the pace and pattern of normal human activity in any remote corner of the globe. In a way, the 'self' is now lost out as in case of the suicide-bomber. But the intent of 'conflict' is seen transferred onto the terrorist organization, out seeking his prey (the 'other' depriving his "inviolable" right). Impact-wise, a terror-strike has perhaps as much in-built game-changing integrals, as the most widespread World War II. Indeed, military historians also testify to the said War having been associated with a bi-polar world with the rise of two super powers, consequential 'Iron Curtain' with the US taking the Superpower role, once the Soviet bloc collapsed from its own predilections and self-contradictions.

3

PREVIEW OF CONFLICT IN MANIPUR

That 'conflict'[3] is man-made and thus its recurrence used to be much less in ancient past, when a person moves about a little more freely, without coming into contact (reacting or interacting) with others is self-evident. Even a mere three decades or so earlier, one could safely and securely move about in any remote corner of Manipur even at the dead of night. No tension (lurking fear) of any kind or any fear of being a victim was there. That literally and epistemologically meant that the predatory 'monopoly right to use force' (and public space) lay intact with the State and with none else; even small time thuggery would be hardly ever. [4] The writ of the State used to run large or loom resplendent with the State throughout the length and breadth of the State. As a result any wrong-doer would stand isolated, pinned down and duly sentenced. But today Manipur is passing through a critical period. The wrong-doer gets lost in the crowd and–to cut a long story short–the wrong-doer dares issue even a 'press hand-out' justifying the 'wrong' done as a revenge for some past incident or a (historical) 'hurt sentiment' perpetrated, say, three/four generations ago by the ancestors of the so-called wrong-doer. The polity is thus in some sort of a bind or turmoil, actuated by 'some' in the name of a 'struggle' (conflict), but whose instrumentality is attributed to the 'majority' collectively. The public themselves–including the elite and the media–are not 'free' to deny the 'implication'; they have to accept the alleged instrumentality *(baby as their own),* rather than throw the 'baby' out with the bath water. The freedom of people to speak out–be it consent or dissent from the goings-on–has been muffled, nay choked out, more or less under pain of own security *(read* death).

So precariously are the people posited that a preview of the Manipur situation today reveals that each and every one (native) has to prove one's neutrality. The so-called "public space" is usurped by some unknown (untoward) force, itself the 'outcome' of a conflict-ridden society. Herein comes to mind a stand-offish kind of adage in the credit market, 'the moment you have to prove your creditworthiness, your credit standing was gone.' Truth *per se* is an abstract perception, to be conceived in philosophical abstractions, rather than in real life. Ontologically speaking, any pragmatic representation of truth can hardly be 'neutral'. In other words, 'truth' becomes partial–accessible only when one takes a side, rather than remain neutral and no less universal even after doing so. The side to be taken here in this phase of 'tail spin' of the entire Manipur is that of 'public interest,' which presumes that one and only end of existentialism is 'harmonious and progressive society'.[5]

For instance, 'backwardness' is a reality (or truth) yet spelling out all kinds of imageries and perceptions to different groups of people. Likewise, 'irrational exuberance' among ethnic groups is another factum, which is being interpreted differently by using different prisms by as many identity brigands of ethnic groups. Social insecurity is yet another being looked at diversely by various sections according to their sweetwill, as also the associated emotional integration. The presence of various insurgent outfits has proliferated the complex situation in different layers as if they are interested only in complicating the social milieu, as they are bent on muddying even whatever peace and serenity still surviving in Manipur. As is too well known, insurgency is being supported largely by anti-social elements working as extortionists, although some elements reportedly invest in narcotics *(ganja* or opium) plantation, trade, interstate transportation and even stock market investment, with perhaps no end in sight.

Each insurgent outfit is holed up in compartmentalized subregions of Manipur, having respective frontline organizations, which has not only put the state in a travail of traumas hindering its capability to negotiate with civil societal groups making it too unsure of their responses to voice against violence, as seen in Nagaland where in the face of apparent threats to the media from the militants, and their continued effort to dominate the media, some uncompromising media personnel had to ultimately wilt under relentless attack on their persons and lives. In the process not only the freedom of press was gagged but objectivity in reportings muffled; while more significantly the very sacrosanct 'will' of the government to undertake bold actions mitigated, even if not nipped in the bud, at the peril of losing its own 'majority' in the Assembly through political manoeuvrings of the powerful frontal organizations of these militant groups, unless Nagalim be allowed to be created. So loud and clear was the then message conveyed, add that there will be 'no end to insurgency, if the NSCN (IM)/MPLF are 'not satisfied'. As is commonplace knowledge now, it seriously constricted even the intergovernmental handling of the serious insurgency problem across the porous and highly sensitive inter-State borders of the entire North-East Region. By comparison, Manipur's insurgency problem was trivial.

The year 1972 saw new states carved out of the composite Assam, which led to an almost sudden upsurge of ethnicity awareness in association with lingering (human resource and infrastructure) development in the region. The following decades have also witnessed a bloody era of violence, conflicts (bordering on terror), enhanced influx of Bangladeshi refugees, including Chakmas, further complicated by increased presence of Central Security Forces deployed to combat the menace. Decade after decade it matured from, a visibly small scale and manageable conflagration of unrest to serial proxies of violence and terrorism, of sorts. It is carefully nurtured by UG outfits ranging from stronger to smaller, weaker, inconceivably miniscule proportions to suit different agendas (extortion, thuggery). Without surprise, some of the bigger groups such as Naga insurgents, Bodo insurgents, KYKL, KRA, Kamtapur Liberation Army, ANHV and Tripura Tiger are fully exploited by the ISI. Anyone who has tried to closely follow up these developments with a research acumen can effortlessly point out that what it takes to solve an issue like this does not take all brawn, but involves lot of brain too.

Another queer dimension of the insurgency scenario is its (mushroom) recurrence–out of the ashes of any UG group laying down arms (through ceasefire accord), another rises. This not only negates the *locus standi* of the UG outfit but reveals its sheer lack of single line hierarchy and therefore its unwieldy (self-denying, and non-existential) character. This has been true of the entire region, which could not be easily explained away, other than as a basic antithesis of any UG outfit. In the year 2006, the South Asian Terrorist Portal (SATP) estimated as many as 109 rebel groups in NE India, barring Arunachal Pradesh (where two Naga rebel groups have their presence, either as proactively or otherwise, and also admitted by Union Home Minister P. Chidambaram during his visit in late-March, 2010). Most surprisingly, only a few of these 3-digit outfits are officially proscribed. For instance, of the over 40 outfits in Manipur, between six and eight remained banned till date. Officially many of these are described as 'inactive'.[6] Perhaps, they preempt to remain dormant in the height of CI Ops (difficult to operate), but regroup or revive the outfit from time to time–in the process revealing their vulnerable, flimsy and 'fly-by-day' or urban-based character,–perhaps lying in wait for any formal 'cash-for-peace'. Insurgents may peter out, but never 'insurgency' as an industry (refrain for unemployed) in NE India.

4

SCOPE AND TREATMENT

To study the situational impact of protracted and ever-increasing tension on Security Forces (SF), the Administration and the general public of Manipur; and to recommend pragmatic remedies to reduce stresses so as to revive the passive sociologies of the afflicted areas towards greater positivity; and to facilitate the task of the government in tackling the complicated manifold-layered problems in totality; the subject matter will be treated as below:

Part I: Effects of (Factors) Internal to Manipur and Nagaland;

Part II: Effects of External Factors;

Part III: Splintered Effects of Violence and Tension;

Part IV: Human Rights Violation;

Part V: Suggested Remedies to Restore Permanent and Positive Peace;

Part VI: Conflict Resolution.

5

EFFECTS OF FACTORS: INTERNAL TO MANIPUR AND NAGALAND

As between Manipur and Naga Hills (Nagaland), there has been a recorded close interaction only since the advent onto Manipur of the British in their push towards the "North-East Frontier" when they chose to reach out to Manipur through the North Cachar Hills (Diphu) subregion after developing their commercial (tea-growing) interest– rather than, down south, through the Cachar plains which had been used by the caravan route in the days of yore, and since followed up by the medieval kings of Manipur to proceed to Tripura or even Assam. In particular the services of Gambhir Singh, Maharaja of Manipur (1821 & 1825-34) was obtained by the British to control the rebellious Angami Naga tribes in and around Kohima. On his part, the ever helpful Maharaja Gambhir Singh led the Meitei militia, just as he did against the Khasi uprising earlier before retrieving Manipur from the Burmese control while during his stay in Sylhet was helping the British to construct a road to transport Khasi limestone to build Fort Williams at the then Calcutta. His, since famous, Naga Hills Campaign has been elucidated by James Johnstone[7] to show that Angami and other Naga tribes had respect for king Gambhir Singh, and consequently that any Manipuri used to be respected where a Briton could not move freely for fear of being murdered, so strong was their head-hunting cult. But early 20th century saw the British constructing a new bullock-cart road between

Kohima and Imphal to facilitate greater (rice and handicrafts) trade for onward transmission to Dimapur, the nearest rail-head serving Assam hinterland. Yet all that "respect" *(read* mutual love) soured during and after World War II, perhaps because of "over-exposure" which comes up as an obstacle even between a couple (two loved ones) as just found out by empirical researchers. Happy couples after "over-exposure" while holidaying during the Christmas (2009) vacation–away from their workplaces and limited exposure-at-home–saw divorce petitions (requests) in Western Europe pouring in with an extra 60 percent. Likewise, the Naga expansionist elements probably saw Manipuri ethnic group as a dominant group hence with a bottled-up sentiment (syndrome of suddenly being exploded). For during the post-Independence Naga uprising even Manipur-bound passenger buses plying along the Imphal-Dimapur (National Highway 39) looted, and Meitei passengers recklessly shot with WW-II-discarded weapons on the Kohima-Mao segment of the NH-39.

In particular, educated youth from Mao and Ukhrul–the area bordering on (Kohima area) Naga Hills have gone over to Kohima where for trade and service, which led to great affinity between them. Most probably this affinity later saw Naga Hills perceive of a possibility for the conceived "Nagalim area" to include Manipur's Ukhrul and Mao/Tamenglong region as "South Nagaland." Apparently political exposure came to Naga Hills during the inter-world war years when the Naga Hills recruits (contingent) came back from Europe at the end of WW I. Under the inspirational leadership of the Hutton, the British Naga Hills Deputy Commissioner, they formed the Naga Club, which set out on a "well-thought-out road-map" to evolve "a pidgin Assamese as *Nagamese,* for bringing about an over-arching linguistic semblance of unity" [8] over all heterogenous Ao, Angami, and other different dialect-speaking Nagas. The Naga political pioneers, including those protagonists of separatist movements claim that Nagas were 'independent' and that the Britishers only colonized them during their advent here. The movement came up only with their Four-Point Proposal (1946), which was based on the Naga Club's memorandum submitted to the Simon Commission. In fact, the Naga activists later expanded their original proposal to a Nine-Point Proposal in 1949, which was again expanded into what they called 'Sixteen Points of Agreement' which culminated into the Government of India according creation of a full-fledged (six-teenth) State of the Indian Union in 1963 (inaugurated by the then President Dr S Radhakrishnan in Kohima with great fanfare and flurry of activities).

The right backdrop to the Naga Hills movement is aptly provided by the Government of India in its documentation: *The NSCN Saga of Violence Continues:*

> "Born out of violent opposition of a minority of Naga insurgents opposed to the "Shillong Accord", the NSCN is wedded to violence since inception . . . Isaac Swu, Th Muivah and S Khaplang rebelled against peace and broke away from their mentors to form National Socialist Council for Nagaland in 1980."
>
> "Since its inception, the organization has maintained a philosophy of violence. Its leadership has been rejecting peace initiatives as it believes in 'arms and arms alone.'"
>
> "Ironically, the politics of violence took its own toll when Khaplang expelled Isak Swu and Th Muivah in 1988. In the bitter factional fights that followed, large numbers of activists were killed by both the factions of the NSCN." (Ibid: p23) [9]

This agenda has not been scrupulously defined in either *Proposed basis for discussion* submitted by NSCN on 21 September, 2001 for the Indo-Naga Issue or *Memorandum on the State of Negotiation* submitted by Government of India on 1 April, 2004. Thus any substantive talk during 2005 may well proceed on the agenda of *Federal Relations,* although there is no certainty even about that, because after all the last extension of 'ceasefire' was limited to six months rather than for one year as hitherto, for unexplained reasons, during which there is no knowing if Naga talks come full circle or Nagalim becomes a liability, preventing Naga Hills District from coming to terms with its own history since 1875, and poisoning relations with its neighbours.

On the one hand, NSCN *prima facie* objects to Nagalim remaining within the Indian Constitution. But on the other hand, it has virtually (according to indications available) accepted a position very much approximating to that. Hence it becomes a puzzle, for the curious observers which one is what:

"Karan Thapar: So, on this issue, just to sum up, on this issue of areas outside the present borders of Nagaland, you want a decision in principle recognizing your claim?"

"Th Muivah: Sure."

"Karan Thapar: And then your are prepared to give them a reasonable time frame to fulfil it."

"Th Muivah: If they are prepared."

"Karan Thapar: If they are prepared. And one route that could be acceptable is a state's reorganization or a state's boundary commission, provided it is done sincerely."

"Th Muivah: That is from the Indian side. That is Indian system. But if they can work out which can be acceptable to us, of course we will accept that. But it has to be a political solution."[10]

The NSCN manipulation is reminiscent of the infamous mannerism or stance of the vanquished *Nongban Kongyanba*–the anti-hero in the Manipuri legend, Khamba Thoibi legend–trying to wriggle out even after falling flat on his back in an indigenous wresting game with the victor Khamba above him, that he was in fact the victor *i. e.* after effecting a quick 180° degree turn of his head just to prove his point that his 'face was facing the earth.'

Even if all administrative powers be handed over to Nagalim, four subjects viz: *defence, external affairs, communication* and *currency* might be retained by Government of India. It is just like the Instrument of Accession respecting either Kashmir or Manipur (1947). Even in these excluded subjects, the Naga leadership claims during the peace-talks that Government of India should have earlier consulted them, if at all any decision directly affecting them would have to be taken. If there is an invasion on Nagalim, Government of India should come to the former's help; but that the former would not come to the help of the former, if India is invaded by a third country. They are thus not even on par, with Nagalim claiming one step higher than the Government of India in the 'Indo-Nagalim Federation'. Far from consenting to this NSCN proposal, Government of India has apparently not even perhaps considered it seriously. In fact, the *special federal relationship* (or autonomy) proposal would require a thorough national debate, including consultation with the opposition, because without the consent of the latter, the Government of India do not have the requisite strength in Parliament to amend the Constitution. And at least so far no consensus has emerged on *autonomy*. Ultimately the Tangkhul Naga leadership may stretch their *unique history* paradigm as a peculiar sharing-in of sovereignty between Nagalim and India but the general assessment is otherwise.[11]

Rather, the NSCN has made a U-turn in its half-century old insurgency unleashed just for an Independent Nagaland. For any event there is a cause and a consequence; at least as long as one is fastidious about analyzing an isolated event. Likewise for a concatenate chain of events some logical flow or syntax of consequence becoming a cause for some other event is admissible.[12] On a time scale, cause is thus positioned in the past while consequence in the future; cause is attached to memory, while consequence follows cause. While negotiating on the basis of *unique history* for *Naga sovereignty* Thuingaleng Muivah has encapsulated both history and sovereignty; past and future; cause and consequence, so that it would displace the entire universality of human experience. The region witnessed a bloody era ethnicity upsurge, consequential violence and terrorism further complicated by influx of Bangladeshi and Chakma refugees also confounded the government, thereby hindering its capability to negotiate with militancy. In fact, so many civil society organizations (CSOs) function in Manipur now that it becomes difficult–if ever the government intends to send out feelers for negotiated settlement–which CSO is aligned to what militant group.

In the process, these frontline CSOs have traumatised both governance and the media functioning, even where insurgent outfits do not directly hinder governance. Lives of various key bureaucrats have been at stake, for refusing to pay heavy stakes or remain at their beck and call; as also many offices in and around Imphal, the capital city have been functioning merely namesake.

Nagaland has produced many politicians, and protagonists of separatists movement from 1930 onwards, enabling the present separatist identity brigands to date Naga Independence movement as far back as 1930. In addition, Nagas claim to have a "unique history" and that the final Naga solution lies in unification of the so-called Naga-inhabited areas of Assam, Arunachal Pradesh, Manipur and Myanmar. Admittedly, hill tribes occupy areas in clusters each having distinct antiquity (distinct pre-historicity), demarcability in terms of ranges, rivers, water-falls and at places even jungles and vast uninhabited areas in between acting as buffer areas from tribes to tribes. The distinctly recognizable blocks in Manipur are:[13]

a) Jessami-Chingjaroi to Kachai-Khongdei;

b) Oinam-Maram-Sangkhomei;

c) Makhan to Songsong-Pakhomei-Mao-Tadubi;

d) Ukhrul-Paorei-Tolloi;

e) Shansak-Bugpa, Chassad-Chamu;

f) Shajouba-Purul Oinam

{with an opposite block in Myanmar; seventeen Naga villages with Kuki as Chief-of Somra.} But in contrast, Nagaland villages have some peculiar features:

a) the Konyak tribes in Chen group of villages;

b) the Totak under the Chief of Chui;

c) the Wakching group of villages under the Chief of Wakching;

d) the Chaklangshu and Monyakshu Ukha range under the respective Chiefs;

e) the Tobu, Tamkhong and Jimpang groups ofvillages;

f) the Pangsha, Wenshoi, Noklok, Wui etc;

{with an opposite block of 129 villages of the Khiamnungam/ Khyamnungam tribe in Myanmar.]

g) the Ao-Sema and Sangtam Changs across the river, Dikhu with Chare as gateway to news physicalities.

h) the Phoms of Longleng area merging into Dissoi jungle are a few examples, like Japhukhong Range, Asetkhong Range, Changkikhong Range etc, Semas under separate Ranges.

The above uncontrovertingly depicts an outline of the natural and traditional divisions wherein closer interactions have been in existence for years, but such closer human affinities have no significance for modem political protagonists and the pseudo-historians.[14] Instead these fundamental moorings and day-to-day proclivities are being ignored by these protagonists and activists by taking recourse to an 'imagined' past. In fact, these Nagaland politicians are taking a perceptible 'shelter' in a word (current fad among identity-brigands) viz. "wronged" by history. The obvious implication of such a far-fetched interpretation is that they can find justification for an 'imagined' history of the so-called "Nagalim". The bare fact is that they are overriding the constitutionality and religiosity of an evolved cohesive living and the inter-dependency for day-to-day needs and feeds. More than anything else–now fancied for these ethnic groups–what they require is their daily lifestyles, traditions, and rituals which bring them their kind of trust and confidence in the continuity of their community; polity, group (clan, tribe etc.).

6

MANIPUR'S INHERENT PSYCHE AND OWN COMPULSIONS

Since times immemorial, a series of struggle for supremacy[15] kept on taking place among the several sections, tribes and clans of Mongoloid people having moorings and descent from both inside and outside Manipur, as true of any area on the migratory (ancient Silk) route–disseminating information, culture and artifacts (goods), and finally ethnicity. Some accounts might register these ingress as invasion or carrying away of property (devastation) or of people (depopulation). But the transcendental fact is that its population never remained isolated from, or un-mixed by neighbouring ethnicities from both the East and the West, whence different "hordes" of attackers came and marauded Manipur. "With passage of time, a mosaic nature *(character)* of the history of the people of Manipur emerged because of alien bloods that had been mixed up in course of time because of "inter-clan fusion".[16] The residual people came to be known as 'Manipuris' and then naturally divided over time into two distinct groups depending on the eco-environs of their habitat: the *Tammis (Meiteis) or* those who came down from the hills and settled into the valley; and the *Chingmis (Haos),* i.e. the hill tribes who remained (settled) in the hills, consisting of different over 36 small groups, speaking different dialects of their own. Over the millennia Manipur remained isolated (landlocked) for outsiders. Religion-wise, by mid-seventeenth century the valley-people adopted Vaishnavism, while three centuries or so later, the hill people–who were earlier animists–became Christians as a result of the colonial administration which allowing the Christian missionary to commence their activities in the hills of Manipur, otherwise kept aloof and un-interfered by the British administrators. In course of time, two distinct religious groups propped up which gradually evolved into two ethnocentric cultures–'one' Westernized and the 'other' Indianized but with a conscious and deep sense of pride in their own ethnicities primordial. In fact the British had an altogether different plan to carve out a Christian state, in North East India as their toe-hold along the Indian Ocean as Chittagong's hinterland, more or less like Singapore, Hong Kong etc, even after they are required to quit India: Even this self-effacing 'fiction' proved a "myth" much to the late realization of some British administrators directly handling North East India:[17]

> "If I were to make to criticism on hill administration, they would be that, in the past, there was too much of a tendency to treat the tribes as subjects of preservation and study, rather than objects for our help and guidance, and we were too chary of extending our administration and too slow in building up self-government institutions among them."

The following major anomalies cropped up and remained not only unresolved over the six decades since Independence but becoming increasingly complicated by a serial competitive arousal of ethnic aspirations which has as on date proved a stumbling block to any possible social assimilation process, thereby keeping at bay any process of cohesion in its queer geo-environmental economy. Some fundamental or core issues are identified below which however require immediate attention for expeditious redressal:

a) Through century-long evolution in semi-isolated ecology, the demography of Manipur has quite naturally adapted to its twin physical features of hills and valley as two distinct groups. The valley which is, as it were, a hill-girdled island of just 10 per cent of the State area naturally with a much higher population of 14.11 lakh persons (or density of 631 persons per sq km against a mere 44 in the hills). Obviously, the five sparsely-populated hill districts: Churachandpur, Tamenglong, Senapati, Ukhrul and Chandel comprise around 90 per cent of the total State area (22,327 sq km) encircling the densely populated tiny valley at the epicenter of Manipur.

b) The hill-inhabitants are not barred to own land and settle in the valley, as and when they become educated and employed in government service through a system of reservation quota for both education and employment. But the non-hill people are barred to own land in the hill areas, even where they stay in the hills during employment, trade or otherwise. This queer 'exclusive' character has persisted ever since the colonial-days, at a disadvantage of the non-hills people. Putting succinctly, some kind of a 'handicap race' is envisaged: the hills people are put at an advantageous position of 'inclusive growth' in what be called the 'mainstream' portion. But not reversibly for the mainstream (non-hills) people, who are barred to own any land in the hills as that will place tribals at a disadvantage in their own environs.

c) Even the education and job reservation quotas extended to as many (the entire) as hill people would seek the same all through the post-Independence era, whereas only SC (Other Backward Classes) become admissible to such privileges. And in a secular nation, that is India, their becoming Christians is not a bar. It must be mentioned here and now that Western education, provided to them has generated increased awareness for changed lifestyle, dress, behaviour and attitude, although their preference for (fixation to) tradition: like ancient rites and rituals, and tribal nomenclatures or identities did not exhibit similar changes during all these years. Their mass adoption of Christianity has brought along all that (social) change, but their deep religiosity (regular church-going) has kept them more or less stuck to tradition rather than to secularity (modernity) through scientific outlook, in readiness for the post-modernity.

d) The Manipur Land Reforms and Revenue Act (MLR & R Act) was not extended to the hills because of constitutional constraints, which were enshrined therein to protect land, which is the only resource on which tribals subsist (incidentally on primitive methods of farming called 'jhumming' which depends on much less labour-requiring 'slash-and-burn'. But there are yet other associated (consequential) hindrances. For, the Panchayeti Raj Act–seeking socio-economic empowerment and institutional reformation or Mahatma Gandhi's *Gram Swarajya* concept–cannot be extended to the hill districts, where each village has its own village *Khullakpas* (or chiefs, or headmen or Rajas). If the Schedule VI provisions of the Indian Constitution is extended to the hill areas–as now aired by some leaders–most tribal leaders admit their (privileged) education jurisdiction will become curtailed to Class VIII institutions only, hence all the higher education needs, thereabove, divested. If the MLR & R Act is not extended to the Hills, land remains un-surveyed and no *jhummia* can obtain individual *patta* (land-entitlement proof) for any hill plot, thereby debarring the bankability of his hill plot for even viable hill development projects and schemes. Herein, the recent developments in the Indo-Myanmarese town of Moreh (on the Indian side) is a case in study, because of border trade, proving a driver for change in its multi-ethnic demography.

e) Most of the disputes in hill areas, meanwhile, among the Kukis, Hmars, Gorkhas etc. embody another big constraint for any possible development project in the hills. These are mostly either directly or indirectly land-based disputes for control of its yield and thus become protracted. To make the confusion most confounded, each tribe will have their own customs and traditions in matters relating to land and settlement of disputes, although generally there is some sort of uniformity of these customs and traditions among, say, Naga tribes, as among Kuki tribes with different identities and nomenclatures.

f) Increasing animosity amidst competitive awareness for greater amenities, infrastructural development and yet other needs among rival groups has left behind an overall "trust deficit" among the tribes, thus defying any normal (cohesive) solution to the ethnicity-related problems almost toxically polluting the psyche of the otherwise discontented (if not disgruntled) public at large.

g) What may be seen as "public space" in Manipur has hardly seen any out-and-out or forthcoming module of a politician with a very honest and self-sacrificing image. Otherwise, partly to be explained away in terms of its very limited (three-decade-long) full-fledged statehood (or democratic norms and parlance seen only since 1971), at just half of the mainstream background. But in the absence of any such role-model, even the highest-ranked local (government) leader has so far proved just a poor average, with the result that the public has lost all their confidence (trust) in the sacrosanct character of the State machinery. Worthwhile relevance need be given herein to the absence of a strong regional party, itself conspicuous in default more in the hills than in the valley. With the result that no politician will ever plan to 'hang up his boots' (after serving a cherished term for a good cause) but keep on changing (horses) in the electoral battle. In contrast, the clearly discernible advantage in the mainstream political scenario is that there is always some political gamesmanship (one-up-manship) coming up among the crowd of politicians: those out on a missionary zeal to do some yeoman's service to the public cause. The end-result–and far from jumping to the conclusion–is that governance all through has been tardy, or that much of the planned development effort became lop-sided.

h) The ambivalence and vacillations inherent in the demands of the various underground outfits do vary, and rhyme with whatever lacuna hitherto laid out. The insurgency scenario itself is such a mosaic of groups or outfits, rather than one or two clear-cut group fighting for the future of Manipur. Besides very much complicating the environs with negative-oriented violence, ruthless killings, reckless extortion, those hydra-headed insurgency (as many outfits as ethnic groups) the two major groups demand inclusion of sovereignty as a pre-condition for peace talks. Manipur remained typical, and continues to be so even today.

Besides farming, self-employment opportunities are available only in small trade (tertiary sector) whereas some adequately skilled youth are not employed in the IT sector and other private sector avenues. But for the average youth of Manipur, employment is limited only to government sector, and in the face of a negligible scope for upward mobility in the service most educated youth remain unemployed. Ultimately therefore the psyche of the upcoming youth seems to depend more on luck. The average Manipuri relies more on fortune favouring him, rather than himself proving much more deserving, through reskilling or harder work. If a youth hopes more on luck in examination; and even when mature, still more and more on luck while seeking employment rather than prepare harder to outwit other competitors; and while ageing eat, drink, enjoy rather than eat and drink with moderation. One educated unemployed out of every five registrants in Employment Exchanges of Manipur crave for some form of indulgence in Manipur. Such bottom line behaviour is borne out by the desperados forming militant outfits in Manipur, which have over the last thirty years multiplied from just 2/3 outfits to 60 as annexed in Table 1: Intrinsic Conflict with their last known strength and No. of weapons. Deprivation-caused militancy is equally high in Assam with just 42, but lower in Nagaland (7); Arunachal Pradesh (6) Tripura (4); Mizoram (2); and Meghalaya (3). Please also refer to Table 1 A (i) to (iv) to follow VBIG organisations and Table 1 B for KNO & UPF components.

7

EFFECTS OF EXTERNAL FACTORS

Landlocked and hill-girdled Manipur valley may look quite safe and secure from extrinsic factors. Yet it is a travesty of history that it was hardly ever left alone even in the ancient days by either religious preachers (precursors of the current Buddhist circuit) and other mainstream travellers–not to forget the curious pedlars from nearby hills. Without doubt Manipur was known as the important stop to hang their boots and rest their tired legs awhile till they become fresh to proceed onto the Irawadi valley, and beyond to Laos and other Pacific coast monarchies, better known then as *"Suvarnabhoomi"*. This was precisely because the hills and vales were not very difficult for trekkers, or the elephant and horses in the caravan to negotiate and also because, as already stated, Manipur valley proved a hospitable land where strangers would find a ready host.

Such connectivity and orientation of Manipur continue to hold out till date as the entire stretch of the Indo-Burmese boundary (between Manipur and Burma) is still not only un-fenced but also undemarcated with people freely crossing the supposed boundary, with neither any security outpost dotted along or within even a few kilometers of the boundary. And the Burmese area just across Manipur limits has offered a safe recluse even in fresh memory of the present population for Hijam Irabot Singh, a Manipuri rebel in the early post-Independence days (who spent his last few days inside Burma) and many other communist-rebels. The apparent 'bind' of this vulnerable border on

Manipur has proved so much more suffocating ever since the above cited Manipur-based outfits found their safe haven over there. According to sources these outfits are much more at home there in the practically unadministered northern Burma area lying just across the Patkai Range, as also all over the Arakan. In fact, adding fuel to fire are those Burma-based outfits annexed in Table 2. These Burma-based outfits often cross over to the Manipur side again for their operations, at a safe distance from the hard-nosed junta military operatives, although much more tolerant to Manipuri outfits.

There are two categories of Myanmar-based outfits: as many as 17 (seventeen) in category A (Existing in Ceasefire Agreement with the SPDC); and 7 (seven) major + 4 (four) minor outfits in category B (Operating outside any ceasefire agreement with SPDC. Unsurprisingly, one finds two striking facts: first, that some of these Myanmarese outfits are themselves either rooted on Thailand towards the east or on Mizoram towards their west (not to overlook the solitary Karen National Union supported by the US and Thailand); second, that some of the minor outfits apparently of recent origin in Category B are supported by Muslim countries. Worthwhile mention need be made of the fact that this outfit-infested area of Myanmar is quite close to the so-called Golden Triangle at the trijunction of China, Myanmar and Thailand. To that extent Myanmar seems to run parallel to Afghanistan's initial 'flash-point' instance of narco-terrorism. But unlike Afghanistan, Myanmar may very well be a poor case without Pakistan and the contextual 'imponderables' in the Middle East. In fact, the real antidote to the area supporting narco-terrorism will be China's 'globalization outlook'[18] in league with that of India and of the other South-East nations. In fact, Muslim-outfits are found active in peripheral areas: on its extremity and on the seashore, rather than deep inland.

8

MYANMARESE FACTORS EFFECTING MANIPUR

Manipur at the easternmost periphery of India did prove better-connected with Myanmar (erstwhile Burma) than (towards the west) with the mainstream, as evidenced by millennia-long contact and association. Written history records the earliest instance of a shared-in experience (a joint campaign: A D 1472) of the Meitei king Kiyamba and the Pong (Burmese) king, Choupha Khekhombajoint campaign's. At times the friendship soured and turned into hostility, reprisals, and the relationship went on till the 19th century when the Treaty of Yandabo 1834 obtained recognition of king Gambhir Singh's Manipur monarchy from the Burmese enmity once for all, although at a heavy cost: viz. cession of Manipur's Kabo Valley domain, more than twice the size of Manipur. Putting succinctly, the relation has been one of 'leaps and bounds'–i.e. both high and low tides of a shared-in millennia-long history. In fact the shared history was itself even-handed, as it was between two "oasis", each surrounded by "barbarities," to recall what C J Lyall: [19]

> " . . . I have always taken a lively interest in this singular oasis of comparative civilization and organized society, set in the midst of a congeries of barbarous peoples, over whom its rulers exercise an authority which, if scarcely approaching the settled policy of more advanced communities, is at least in the direction of peace and order.

> The valley of Manipur in several respects resembles in miniature its neighbour, that of Irawadi. In both the civilized people who occupy the central settled and organized region are nearly akin to the wild folk who inhabit the hills which enclose the alluvial plain. But while Burma has accepted the mild and gentle religion of Buddha, and this profoundly modified the original animistic cult, Manipur has been taken into the pale of Hinduism, and has imposed upon itself burdensome restrictions of caste and ritual from which its greater neighbour is happily free."

The post-colonial era did not see Burma thriving long but soon went into some form of political transition. More or less in that "political flux" went in all the South-East Asian countries: Myanmar, Thailand, Malaysia, Indonesia and Philippines. The 'defining political aspirations' in each of these countries has been distinctive in the sense that they with yet others much earlier formed the Association of South East Asian Nations (ASEAN) apparently aimed at forming a trade and economic grouping, but over all these two decades or so failed to evidence the much needed "integrative" attitude to one another. As a result the Asean remained almost nominally operative, in the hope of picking up momentum as and when an "overarching trend of transition" onto the regional aspirations towards some common goalposts, such as a set of norms for governance within state boundaries. Also absent in a linkage among mutually exclusive activist-groups which are seeking, along parallel tracks, a new political ambience in their respective domains. Of these, Myanmar in particular with their long contemporary experience of military dictatorship, still remains unexposed to democratic norms (or consensus decision-making)–and even the 2010 election, abstained by National League for Democracy (NLD), led by the still incarcerated Nobel Peace Laureate, Aung San Suu Kyi. Priding over this, the State Peace and Development Council (SPDC), as the Myanmar junta styles itself, will complete the task of "restoring-democracy." Hence, the political flux is likely to remain for quite some time, at the peril of any immediate return to democracy.

Chinese assistance to Myanmar infrastructure on the Manipur border "triggers Myanmarese trade boom." A bridge has been constructed over river Mynal just at the Myanmar border towards

Manipur, much to the benefit of Myanmarese traders coming over for border trade at Moreh. In sharp contrast, on the opposite road (NH 39 or the Imphal-Moreh stretch) there are no visible signs of improvement, which spells out a definite contrast, seemingly between Chinese influences reaching out to India, via Manipur. In fact, the Myanmarese side of the border is in sharp contrast with the Indian side has well-constructed roads, supermarket facilities, and neat houses along the road leading to Kalewa from the border town of Tamu on the Myanmar side, just opposite Moreh (India's border town). Besides defining the Indo-Myanmar international border, the construction of the bridge has provided a short-cut over Myanmar area for the Myanmar traders while coming to the Namphalong market, by avoiding the circuitous road passing through the porous Indian land on the Manipur side. These traders come to the border market for daily two-way trade: selling their own local produce and buying back their required articles from the Indian side. Needless to add, the same traders carry on illicit trade viz. selling cheap Chinese electronics, silk goods and a wide range of goods, so much so that China has become the net beneficiary of the back channel of the Indo-Myanmarese border trade, now being carried out through Gate No 1 as per an agreement signed way back in . . ., whereto further effort is directed to boost the border trade by extending the trading activities at Gate No 2 as well. After completion of the China-assisted bridge, the border trade has not only boosted the border hinterland on the Myanmar side, but also opened a large market for Chinese electronics and other manufactures, much sought after and smuggled in across Manipur and NE India even over to mainland.

Mention herein need be made of the fact that this very Moreh-Imphal stretch in NH-39 happens to be part of the much-vaunted Tokyo-Istanbul Asian Highway (AH) No.1 so as to ultimately connect with Mae Sot in Thailand via Bagan in Myanmar. Alongside the AH should ultimately pass the Trans-Asian Railway (TAR) and its transport-cum-warehousing infrastructures when it ultimately comes up. Also, the upcoming East-West or the Saurashtra-Silchar Super Highway would take on the AH further across Bangladesh to mainland India, thereby overcoming all the handicap resulting from the Partition (1947). In fact, on the Indian side even the Lokchao bridge on the Indian side of the border a little further inland has not been repaired ever since it was constructed during the Second World War. But the biggest explanation for the contrasting scenario that side of Myanmar is due to the military junta having been comparatively been very swift in the pace of decision and execution of development works, due to the commandeering effect (by way of public participation), unlike democratic India. Of course, an improved Indo-Myanmar diplomatic relation has also seen India's Border Road Task Force (BRTF) having crossed over inside Myanmar and already completing construction of the "Friendship Road" to Mandalay from Moreh (the point of the bridge across Mynal river at the international border), with the projected possibility of plying regular passenger buses between Imphal and Mandalay in the first instance, hopefully opening a Buddhist tourist circuit from across Japan, China, Korea, Thailand and Myanmar extending right upto Gaya in east Indian mainland, as an alternative of the currently air-borne circuit.

According to historical sources, there are great expectations forthcoming from both the sides to bring about a change in the ongoing character of border trade, including the raising of the volume of border trade activities from the present 22 to 44 items through this road stretch. In fact, the very trade-handling infrastructure in Moreh border is in for an overhaul shortly. To bring about such a change in the character of the ongoing border trade and speed of the import-export activities throughout the country, the Land-Ports Authority of India had already approved the setting up of an Integrated Check Post (ICP) at 13 different locations along the border with Bangladesh, Nepal, Pakistan and Myanmar. Towards that end land acquisition process in Moreh is over and 45.50 acres of land has already been acquired from 405 land-owners, and the Detailed Project Report (DPR) of the proposed ICP has also been finalized to start construction at an estimated cost of Rs. 130 crore modelled after the Wagah border in Punjab, prepared by the Rail India Technical Economic Services (RITES). When completed, the facility will provide facilities for land customs, police, customs prevention department, forest, narcotics, postal, banking, telecom, animal and plant quarantine, quality certification inspection agency, trade facilitation counter, food testing laboratory, parking and cafeteria.

On date India at the national level has conveniently disregarded its close military and economic ties with China making the latter an important player in the Indian Ocean strategy (including Myanmar's junta assistance sought from China for oil and gas transportation across the Bay of Bengal, thereby avoiding the piracy-prone Bay of Malacca, in addition to developing five new ports: Putao, Paungbyin, Mandalay, Pegu and Yangon, right from the Victoria Point in the south to Sittwe (almost at the southern tip of Mizoram) in extreme north along the coastline (where India has been planning a sea outlet from Mizoram through a river); and also setting up radar station, airstrip and naval base at the (former British) Cocos Island, where ships carrying out clandestine duties of unloading insurgents, armaments for onward transmission by boat to Kanthalang on the Mizoram border on their inland trekking to Manipur, Nagaland and North Cachar Hills in Assam. Mention need be made of their trekking up to Tipaimukh—whose proposed dam and associated township—would intercept the UG route northward.

9

NON-MYANMARESE FACTORS EFFECTING MANIPUR

As a correlate and also as a touchstone, China looms largest among India's eastern neighbours, barring Bangladesh and Bhutan in the West, of which particularly the former seem to come closer to sharing commonality with India. In fact, Chinese influence transgresses throughout Myanmar in whose road-construction, naval and defence interest China has been long committed, much to the chagrin of India, except during last few years when Myanmar seem to start welcoming Indian interest to some extent. (..........)

Farther east, Thailand is no stranger to coup-masters or to the political masters of the armed forces. Even now the Thai form of democracy seem to be monarchy-inspired, with the Thai king as the head of the State, and still very much revered as even more than in Japan, and the king with an unusual halo still acting as some sort of catalytic agent. The challenge for the military-friendly Prime Minister Abhisit Vejjajiva (since 2007 when coup-masters held a general election) is how to thwart Thaksin Shinawatra, the twice-elected prime minister, now on a self-exile following a judicial verdict against him. With Thaksin-supporters still ganging up protest, he had to even declare an emergency in March 2010, it still looks difficult for Abhisit to deplug Thaksin, his sworn political rival, from Thai politics. The only preponderance is its southern parts which are dominated by Islamic fundamentalists yet needing both articulation and careful underplaying of concerns in India's foreign policy stance and outlook in general and Look East Policy in particular.

Of the five South East Asian nations now in a political flux, Malaysia is no exception. Although still robust-looking in successfully keeping away military away from democratic practices so far, Prime Minister, Najib Tun Razak has announced his "New Economic Model" (NEM) just to render equitable benefit to the three major groups: (a) Bumiputras (Muslim majority + indigenous population), (b) ethnic Chinese minority; and (c) people of Indian origin, another significant minority group. In this "upscale" developing economy inequality is being addressed quite rightly through Tun's affirmative action, rather than through the more fundamental calculus of basic statehood and statecraft. One needs be wary of her relationship with China in the context of the past as well as the emerging situation in the South China sea, yet its foreign policy has inwardly emphasized better relationship with its Asean partners generally and the Muslim world in particular. Although quite friendly with India and appreciative of its security compulsions, it innately expects India to act as a counterpoise to the perceived regional threat from China. Policy-scientists need thus be wary of the growing fundamentalist influence in her polity, lest it affect India's growing interest and policy-parameters.

Right from the Indo-Pak conflict (1965) onwards Indonesia was supportive of Pakistan by offering naval base, and even claimed renaming the Indian Ocean as the Indonesian Ocean, although later on because of the engulfing East Asian Economic Crisis and yet other separatist insurgencies, it had to be more inward-looking, if not for anything else, but for need to immediately consolidate national identity. Aside, fears of the insurgency in the troubled province of Aceh and its spillover prompted joint naval patrols (with Indian navy) in the Andaman & Nicobar Command (INC) seas. Yet Aceh remained a fertile breeding ground for fundamentalism, with reportings about presence of Al Qaeda operatives (a possible sanctuary) in Aceh. One may introspect, "If Al Qaeda is there, can the ISI be far away?"[20] Indonesia looks equally promising under President Susilo Bambang Yudhoyono (SBY), having transited from his erstwhile addiction to the praetorian model of governance. Having been recently re-elected with an authentically-conducted comfortable majority he looks set, as based on praise in regional circles for this particular success story. Besides, the military is no longer breathing down civilian authority's neck, as true of others: Myanmar, Thailand, Malaysia. Vigorous anti-terror operations and drives have been carried out, besides persistent anti-corruption raids (perhaps benefitting his party coffers).

Likewise, politically stable Philippines is still running high on the glories of the efficacy of "people's power", against autocratic rule in the past. Of all the controversies during Gloria Macapagal Arroyo's long rule (due to be over till last election in May 2010), the "political legitimacy" of her ascent to, and stay at the helm has not been fully sorted out. In that sense, the political flux persists.

10

'OTHER' FOREIGN POLICY ISSUES

Insurgency in NE India in particular will stand to gain from 'mutual trust' or a "genuine friendship" between India and China, the two most populous nations in the world, according to Nobel Peace-laureate Dalai Lama in Washington during his private visit (late-February-2010) to the US to meet President Barack Obama and US Secretary of State Hillary Clinton. What the ageing Tibetan leader is expecting from the Chinese leadership is to develop mutual trust, which can be achieved by becoming an "open society". If it wishes to realize its goal of becoming a superpower, China should lay emphasis on moral authority by ending censorship. Only then Dalai Lama said mutual trust will develop between India and China. Contending that censorship and hiding the truth are 'immoral', the Tibetan leader said the Chinese people had a right to know the reality: "And also, you see, the Chinese people have brains to judge which is right and which is wrong. So there is no reason to put lot of control about media. That's wrong. This I really feel very strongly." This internationally respected religious leader is "fully committed" for Tibet to remain a part of China, but sought "meaningful autonomy" to Tibetan autonomy. Even the US officials commenting to the press– after Dalai Lama met the US president much against protestations and warning that the said meet will be deemed by China to seriously damage Sino-American ties. China's huge problem lies in its increasing behaviour as a rival point in Asia protesting sales of US arms to Taiwan, Washington criticizing China's valuation of its currency (despite its accumulated trade surplus of around $ 2 trillion) and the cyber attacks on Googles, which the US claims originated from China. January 2010 saw China suspending military exchanges with the US following Washington's announcement of $ 6.4 billion arms sales to Taiwan, including Patriot missiles and Black Hawk helicopters. Whereas China is trying to increase its sphere of influence not only in Asia (containing India), the US is trying to likewise keep India through strategic nuclear help for peaceful uses.[21]

11

CULTURAL POLARIZATION IN MANIPUR

Unlike the encircling terrain of Manipur inhabited by sturdy *jhummias* (tribals), the Imphal valley even in the very early age proved a convenient resting place for preachers, travellers and pedlars passing through the Manipur valley over to Burma *(Suvarnabhoomi)*, Yunnan (south China) or beyond. Over centuries, it acted as a melting pot of diverse culture whose outcome transcended as a syncretized culture, of course underpinned by the hybrid character of land-owner-farming valley population with settled lifeways.[22] Such an ethnic genesis of Manipur offers the best starting point for cultural polarization. This view on the peopling of Manipur–including those of this contributor[23]–is in sync with that of Gangmumei Kamei:[24]

> "The people of Manipur consisting of three major groups, the Meiteis, the Nagas and the Kuki-Chins are strongly influenced by ethnicity and primordial consciousness. There is a strong search for their ethnic genesis, their ethnic roots and origin. While the Meiteis are a nationality with a long history of nationhood, the Nagas and the Kuki-Chins are ethnic entities marching towards pre-State to non-territorial nationality. Manipur is a plural society. The population is predominantly Mongoloid with some non-Mongoloid elements, the Aryans and the Tai who speak the Tibeto-Burman languages."

The ethnicity difference (or genetic distance) between the Meiteis and hill people are only in terms of degree. Even if this is conceded, faith-wise, there remains a mystery: as to how even the pre-Hindu faith of the valley population could be out of sync with the almost purely animistic character of the hill counterpart; and also why the 18th century Hindu kings did not extend Hinduism to the hill people, if at all considered seriously? These pinpricks notwithstanding, the historical fact is that the hill-plain fissure, if any during the ancient days, became looser and that could very much explain the hurt sentiments, mistakenly perceived by hill counterparts when the Vaishnava Meiteis started adhering their "do's and dont's": including avoidance of any contact between the sanctified Hindu and 'others'.

The most plausible explanation Recalling Manipur history, one has to admit that, although land-locked Manipur valley itself has been frequented since the ancient Silk route days by various exotic people, while natives of both hills and the valley of Manipur have been crisscrossing outside Manipur both to the east and the west of Manipur either as cavalry or infantry of the kings during their adventurous exploits, whereby also Manipur saw exotic elements and genes. Then again the colonial era saw further interest and direct contact between natives and British officers, their escorting soldiers, and other henchmen. No doubt, the Anglo-Manipur Treaty of Chittagong (1762) between Maharaja Gourashyam and Verelst of East India Company heralded a formal opening of trade and economic contacts. Although that treaty itself proved infructuous, as the innovated technique of column-invasion–most effectively used by the British on the Indian Plains–proved a tactical disaster in the face of steep gradients across multiple ranges of hills from Silchar to the Manipur valley.[25]

Then the *Chahi-Taret Khuntakpa* (1819-26) or Seven Years' Devastation of Manipur valley further saw the annihilation of Manipur militia and their occupation of Manipur valley for seven long years during which Manipuri women became exposed to Burmese soldiers' advances. Also while retreating the Burmese carried away long prisoners of war, both children and women, to populate Burma, so that in most part of Burma (Myanmar) till now Meitei villages survive, whose services were availed by the Burmese kings as artisans, silk weavers, cavaliers (for the much-dreaded *Arambai* or arrow-heads shot from slings), foot soldiers. Both before and after the Seven Years Devastation, history records various exploits of kings leading victorious campaigns to the east (Burma) as well as to the west (Cachar, Ahom, Tripura). Without surprise, therefore one could most naturally expect a poly-ethnic population in Manipur. For that matter, Chinese masons who came over to Manipur during the medieval era were allowed to settle in Manipur according to ethnographic records of that period, since compiled by British officials assigned to work in Manipur. Of course, Khagemba (1652-66) was generous enough to allow one thousand Muslim foot-soldiers of the Ahom king, who unsuccessfully invaded Manipur to settle down in Manipur valley by granting cultivable land and local girls.[26] In fact, 17th and 18th century Manipur became the destination of many upper caste Hindus fleeing Bengal, Orissa, and UP regions during Muslim rule, and settling in Manipur valley. In other words, the late medieval Manipur saw an emerging polyethnic peopling of Manipur. The king's policy plank of "plasticity of culture" did not come by way of an one-off (exceptional) instance shown towards the defeated soldiers numbering about one thousand. By nature all kings would be considerate to both the valley and hill people. On ground of language, religion difference, they would neither discriminate, nor allow people to feel "deprived" by class distinctions or by poverty. [27]

For the sake of keeping himself informed (connectivity) the monarchy had a regular network of informers (runners and cavalry) so that he could at his own initiative (and waiting for interested parties to seek his attention). What in effect monarchs did was to manage competition and avoid conflict among outlying groups. The correlates of such attitude of the Meitei king to the hard-working jhummias used to be: (a) during peace-time was to encourage tribals to compete but by way of a consistent policy framework avoid conflict; (b) during threat of conflict or war, to make their help good helping the king fight throughout the relevant period of Manipur history. Putting differently, it is in hill people's own overall interest to come to the strategic assistance of the Meitei king in adversity. Thus the valley-hill relationship was well beyond the narrow military paradigm of the king's militia *(lallup)*, enforced in the valley or with full commitment "up to the hill". Above all, the commonality of interests is attached core significance. Such a built-in process of and effective service (company + fighting) in the royal campaigns, with this difference that should the king be desirous of advancing to the West, naturally the Kabui (now re-designated as Zeliangrong) would come handy; if the royal militia proceeds Eastward the Tangkhuls come handy.

For that matter the area beyond Moreh, famously known as Kabo valley (controlled by a system of Thanas at Tamu and elsewhere) was acquired by the king of Manipur during these tumultuous days in which Manipur has been ruled by the great kings of Manipur like Garibniwaz, Bhagyachandra and Gambhir Singh. In fact, one often wonders if (both the history and peopling of) Manipur would have been altogether different, had king Gambhir Singh not solicited British help to retrieve Manipur in 1834, by whose consequential Treaty of Yandabo (1834) drawn up between Awa (Burma) and the British, Manipur was recognized as an independent kingdom {though it ceded the (Manipur's) Kabo valley to Burma}. Naturally all these cross-currents of historical events evidently led to genetic mixing as between various ethnic groups of the peoples of areas now known as NE India and SE Asia.

In the year 1834, Britain assumed *de facto* control over the Manipur affairs in the name of maintaining the Manipur Levy which helped the fugitive king Gambhir Singh to retrieve Manipur from the Burmese hands–although Britain assumed *de jure* or direct control only 57 years later in 1891, after storming Kangla. Significantly enough this period saw Major (later Colonel) McCulloch (1844-62 and 1863-67) working as Political Agent in Manipur, when he married a local lady from Raja Nara Singh's family, [28] who stayed in Manipur for the longest period of 27 years in medieval Manipur (counting his tenure as Assistant Political Agent) and who left behind the most exhaustive presentment of the State and its peoples in the Account of the Valley of Munnipore and of the Hill Tribes, based on his intimate knowledge of Manipur, whereas Captain Gordon of the Manipur Levy compiled a dictionary of English and Manipuri.

With an exceptional sense of adroitness and liberality in thought and action–given the overall primordiality cover all over the North-East Frontier–Manipuri kings would hardly show systemic indifference to the ethnic (hill-resident) minorities: even *vis-a-vis* the valley-resident minority, Manipuri Muslims. In case of a tension build-up between two hill villages arising from, say, 'hot pursuit' of game, the king would send out (or lead) militia, but more to bring them to 'peace and order', whereupon as were the custom, retributions collected, which was paid up more as compensation for harm done, but without any discrimination. And the post-conflict 'merit-feasting' is most respectfully hosted by the offender, as in case of king Gambhir Singh's taming the Naga uprising at the instance of the British. Or else, the king would hardly if ever interfere in tribal (hill village) affairs. Yes! Perhaps 'Back to basics' (or gun culture), if erring tribal chief is found defaulting in paying up regular tributes and other dues. In that age at least, Meitei kings' actions used to exhibit orderliness and peaceful self-respect and amity and adherence to the "settled policy of more advanced communities. As of 1881 C J Lyall found in the Imphal valley "this singular oasis of comparative civilization and organized society, set in the midst of a congeries of barbarous peoples over whom its rulers exercise an authority which, if scarcely approaching the settled policy of more advanced communities, is at least in the direction of peace and order". [29]

12

THE NAGA TANGLE

On ultimate analysis, two extrinsic and altogether independent historical factors can be attributed as to have come together and opened up (over-exposed to the wide world) the otherwise land-locked and tradition-bound Manipur-Naga Hills area enough to cause the so-called 'Naga tangle', and subsequently insurgency throughout the Northeast region. But for such sudden and all-out exposure social scientist doubt, if the natives could have changed their psychosis or taken a U-turn in their attitude, temper, mobility and adaptability to varying environs. By now it is common-place knowledge that the Allied Forces, under the overall command of Field Marshal Slim at his Kangla Hqs, were 'desperately' fighting in Kohima and Imphal to prevent the advancing Nipponese and Indian National Army (INA) from getting a toehold in Manipur and Naga Hills, which led to modem warfare in all its most horrendous forms: hand-to-hand fight, aerial bombing, aerial dog-fights and all those devastations. Of course, military historians attribute '3-M' (monsoon, malnutrition and malaria) for the Allies' victory, later hastened by scientific advances as embodied by the atom bombing of the Japanese cities of Hiroshima and Nagasaki (6th and 9th Aug 1945). Naturally, the retreating armies discarded weapons which indirectly added to the woes of the civilian establishment, itself dislocated and weakened by war. These left-over weapons in the Naga Hills, Manipur and also Patkai range from the retreating Japanese forces, the INA, including some even discarded by the Allied Forces in World War II. In particular, INA took help of natives for both information, food supplies and other strategic assistance, which raised (if not tilted) their sensitive (freedom) perception, apart from contributing to a sudden, rather than, a gradual transformation of their narrow (existentialist) perceptions and traditions towards modern warfare, scientific advances, in short, modernization.

Apart from these weapons and other discarded military hardware (jeep bodies, engines, vehicles, which became accessible to those Naga leaders, whose identity and ethnic aspirations have been awakened by the British officers. Meanwhile, some significant developments have taken place in the Asian region. While the emerging Pacific nation, viz. the US has been occupied with the Far East (monitoring war-rehabilitation in Japan or fighting in Korea, Viet Nam), Communist China has also emerged, forcing Kuomintang to recede to Taiwan. And a resurgent China has been helping militant groups from the South Manipur, Nagaland and Arakans (Chittagong Hill Tract Area). There are definite information that UGs sent four gangs to Yunnan, South China, through Upper Burma *i.e.* Kachin and Heimi Naga tracts and procured automatic weapons at least upto 1975. And these first few batches of the Naga Federal Government (NFG) militants had to encounter vagaries of nature like inclement weather, arduous and circuitous trekking through dense jungles. Soon enough the NFG slogan of "Nagaland for Christ" saw the clash of religious ideologies. And the consequent forceful intervention by the Naga Baptist Council (NBC) to delink their relationship with China paved the way since 1980 for a new concept in ideologies viz. looting of State and Central SF armouries, ambushing foot and vehicular columns both in Naga Hills (Nagaland) and Manipur hills and valley.

13

SHADOW OF TERRORISM

There is still an odd nascent feeling shared in by think tanks on Manipur's four-decade long conflict situation that insurgents throughout Manipur got the instinct to use force not only from their millennia-long conflicting history–but because their own parents used to relate bedside stories how they saw the invading Nipponese and INA recruits fighting the British and Negro soldiers in and around Imphal; or Manipur valley being bombed again and again, or aerial dog-fighting between the Allied and Japanese fighter aircrafts (then quite a novelty)–as vividly as the by the Western post-war film makers would depict fighting in "Bridge on the River Kwai"–this time for commercial art. Besides these traumatic high-pitched tit-bits, most of them as children used to play with their left-over weapons, scraps of (reusable) war vehicle-components, etc in the dumping grounds in the valley and across the hills. The same would be true of Kohima area in Naga Hills and across in Patkai Range. The fact that the entire NE region remained only nominally governed even during the early post-war years added to the woes ofthe Manipuri people–already exposed to the devastation brought upon by modem warfare. In living memory, Manipur (alive) was in a way broken off most violently from the tradition–then prevailing in this land-locked state–that were the very cultures with a deep-rooted sense of their native ways of feeling, because it was individuals in these cultures who experienced the deepest trauma when the well-knit society broke up their own sense of an organic inner life in this two-millennia old monarchy.

They thus became completely exposed to the twentieth century war modules, materials, scientific modes and temper (and thus to modernity), including the keenly awaited hot war stories among an otherwise detached public of native. Further on, along with this stimulus of newspaper-reading among the literate public sphere came another exposure: requiring greater human needs like higher education–because there were only English high schools but no college then. As to exposure to essential consumer goods, bread, biscuits, tea, soaps, clothing were often just gifted away to the natives who work as contractors and their children who throng military canteen stores and other establishments. But the change brought about in the attitudes, more and behaviour of the people was completely transformational.

Interestingly some characteristics of modernity became self-evident: first, a shift in people's 'sensibilities'; and second, a critical awareness, among people then used to a different lifeways: day-long work, followed by bedside stories for the children from elderly, while the male householders waste evening hours in dull gossip in favourite haunts (hubs) of the village (locality). If the post-war era rudely broke Manipur's tradition, there was a mismatch; since there was no alternative module which would enable their sensibilities to shift to concepts of identity to larger class, gender and regional-linguistic identities and a complex sociological one where the thinking (highly sensible) public leaders (like Hijam Irabot Singh) face the forces of standardization brought upon by the market society and its consumer culture seem to be forming "imagined communities" or "alternative nationhoods" or still others underpinned by pronounced regional, linguistic, ecological and other concerns (including preservation of old vestiges and heritages). Having been made suddenly open to the wide world, people's mindset could not transform to the world as such in the first instance. The 'Manipuri mind' translated the 'world' into 'India', just as it transformed itself with the help of tradition (in the wake of Sanamahi and other revivalist cults. As it were, the 'Manipuri mind' recognized the Other as their own

Self, giving rise to the delusion of duality, conflict and schism. Unfortunately, the 'Other' in people's sensibilities is not the Self's adversary, but the Self's unrecognized aspect–shadow or lingering doubt of survival as a nationhoods (existentialism).

Had there been a 'sense of communion' between the emerging Indian nationhood and the 'imagined' alternative the duality, schism (conflicting alternative) must have vanished. The vital understanding of thriving together through difference seems to have been critically threatened. There are enough reasons to suspect that such analogies–howsoever cryptic *prima facie*–illustrate the interdependent and collaborative (and even synergistic) process of culture Manipur has evolved as its mode of life. Not only in Manipur, but even rest of the NE region this contextual module may hold true, from which conflict-resolution interlocutors may well start any plausible commonality of thought between the *Self* (insurgents) and the *Other* (the mainstream). But then there need be fresh agenda-setting, acceptable to both conflicting parties: unfortunately the same Self and the Other. Any identity idealogue (brigand) will now demand that they talk of history as events unfold in a linear fashion, and therefore that in a basic sense progress is inevitable.

As is well-known, the British ruled through 'divide and rule' among ethnic groups, and through 'metonymy' e.g. psychic repression of Manipur's royalty (generalized) as 'chief' on par with other chiefs. Pithily put, the colonial regime was no less traumatic for Manipur than the holocaust seen during the first global-scale World War. Both represent a history of subjectivity with this difference that if the former's subjectivity was concealed by Britishers' arbitrary choices of metaphors (metonymy) as a site of political gender-struggle over the body. For the sake of record the colonial regime saw as much devastation on property and 'women space' as during the World War II which was standstilled by a nuclear holocaust. For, history has on record a Britisher who could not resist heterosexual/interethnic boundary and married a local princess (which may in fact be metonymically explained away as 'hermaphrodite awareness'. On the whole, one has seen how 'remnants of feudalism' did survive as very powerful psyche in some people, despite the latest newspeak and jargon and a barrage of electronic illusion of informations (data) and a possible delusion of enrichment enough to become one of the two economically powerful global nations in Asia (the other being China). Coincidentally another twist of history is that Manipur (North East India) sandwiches Mainstream India and south China's Yunnan, almost. The intercultural traffic through Manipur and North-East region in the pre-historic India has not only given Manipur a dynamic genetic diversity but has been its secret cultural sustenance for thousands of years. Its biodiversity is expressed in ethnic, linguistic and cultural terms. This made Manipur open even in the 21st century. The Manipuris translated the world into India just as they transformed themselves with the help of tribal traditions. Yet this vital 'understanding of thriving together through differences' seems to have been critically threatened now. Sixty years of freedom with (unsatisfied) needs have pushed that cohesiveness into the wall of geography. May be, some of them are not even in a position to even speak out, far from articulating what are their 'rational human needs' which remain unfulfilled, thereby justifying say their declaration of 'bandh' or 'economic blockade' of the National Highway stretch between Senapati and Kanglatongbi near Imphal. Yes, they know pointedly they are doing it for the sake of forcing the concerned authority to come to them and negotiate across the table. But if the second- or third-rung activists are queried, "Why do you need hazard-free connectivity on the road, whereas immediately across the road they negotiate deep incline on one side and decline on the other?" They are not in a position to justify their resort to 'coercive path' they are not in a position to reason out. Tolerance-wise, they have exhausted all their patience, but no trace of any democratic input–forsaking their erstwhile peaceful existentialistic culture–betrays in their exhibition of conflicting attitude/mores.

On the one hand, the serial 'bandh' over NHs has been prompted by what may be called the "myth of domination" (*cf* 'area domination' exercise of the Indian Army in insurgent-controlled area through road-opening patrol ROP). The tribal chief has given his group to earmark the land and environs around in terms of 'village territory', across (or near) which the NH must be passing. There is a similar tension build up around sites for development of uranium deposits (Meghalaya) or hydel power (Manipur) which also partakes the nature of area domination-interference between natives and government, which can be summed up as 'resource nationalism'. Pithily put, there is another category of human destructive behaviour with conflict as its outcome:

a) International conflicts: World War I & II, cold war, nuclear arms race, terror-strikes from across the border;

b) Conflicts within a society: rich-poor, ethnic groups, resource nationalism;

c) Personal conflicts: between the family/hostel barrack room-mates (ourselves).

Hence the problem of conflict-management morphs into three areas, which form virgin area for even police (Security personnel), teaching faculties, and the public at large. Hence all these stake-holders need put in their conflict-management effort only through better understanding of the process, roots of human conflict and the involvement of (other) stakeholders, as well as disciplines: such as psychology, sociology, government, economics, and international development. Ultimately, they need also be concerned with four "paths" to resolution of conflict: strength, negotiation, justice, and personal transformation, and the overall strategy to 'lasting and positive peace'. Alongside, the strategy of governance need also be oriented to one of the 'pearl of wisdom', an ounce of prevention is worth a pound of relief.

The complex crisis of culture today is different from those seen in the rampages of fascist regimes both in Europe and Asia in the 20th century. Perhaps, the only commonality lies in the dehumanizing cult, e.g. genocide, or the inter-ethnic rivalry between two religions springing from handed-down holy books.[30]

14

STAGGERING HUMAN RIGHT ISSUES

Militarisation of societies is worrisome not so much because of the threat to life (particularly women and children) and property but because of its flagrant violation of normal human inclinations to ask for their rightful place in society, and freedom to 'live and let live'. This applies to both State actors as well as the commoners; although most seriously deprivative (often coercive) in case of the commoners. Initially well-intentioned to accord 'complete' freedom from a perceived 'enemy', the non-state actors have forced the State actors to become offensive. Military deployment can and need not be always in a safe distance from public places so as to minimize casualty and fall-out on innocent public. More often, the general public has to prove its innocence; and unless and until so established beyond doubt, people are rounded up in markets just to minimise deterrence on public. But very often, it is a 'catch 22' situation even for the Security forces while the insurgents get away after their 'hit-and-run' strikes on targets. As seen in Sri Lanka and elsewhere, outfits even here have never been forced to launch ,an all-out effort 'do-or-die' confrontation (reminiscent of fighting between two sworn enemies as during W W I or II). Well-known are the stories of insurgents seeking one-night refuge in private households, for which the security forces cannot. And the paradox of military operation lies in neutralising the strike power of the insurgents and forcing them onto the (peace dialogue) table, rather than seek a complete annihilation of these obdurate 'sons of the soil'. Underpinning the CI Ops is the ultimate working matrix of a democratic correlates, praxies and principles of governance, which bind even the hands of the Union home minister P Chidambaram from using Army or defence forces in anti-Naxal operations. Nonetheless, such soft comer ('kid-gloving') is not shown to 'terrorist'. May be, this is one of those almost imperceptible points of difference between insurgency and terrorism. *Mutatis mutandi* on the human rights dimension also.

That also precisely explains why the State actors themselves seem to provide a complete range of accommodation to rebel groups ranging from 'cash-for-peace under ceasefire' scheme including political accommodation to Laldenga, the Mizo rebel leader, in the 1980s under prime minister Rajiv Gandhi. That means that there is no hard and fast rule that rebel groups be exterminated, which may also quite unduly cause hurt sentiments. If the rebel leader now in his late prime age could be politically empowered, and the Mizo populace also concurs, nothing should stop the Central leadership to adopt any such 'problem-solving' approach to the best satisfaction of one and all.

15

NAGA INTRANSIGENCE

Will 2010 see a breakthrough (or peace) in the Naga insurgency scenario, now that Prime Minister Dr Manmohan Singh and home minister P. Chidambaram declared (November 2008) their sincere concerns to resolve the Naga issues in the time frame of one or two years? Possibly, a step closer amidst the growing Naga intransigence, and in the face of crucial days ahead to converge on a 'road-map to peace' through a predetermined 'structured dialogue', which does not leave any scope for intransigence (or black hole). But it still looks difficult despite the renewed GOI effort through offer of 'more autonomy' and by entrusting the "interlocutor's responsibility" to an experienced bureaucrat (once serving as the Nagaland chief secretary), apparently with backroom consultation with the NSCN (IM) leadership. Basically because of the faction-fighting between at least two NSCN factions. Fresh in the mind of the Naga populace is the long history of Naga insurgency during which faction-fighting surfaced much to the chagrin of the elite. Since 1997 (when the Indo-Naga Peace talks began) faction-killings went on except perhaps in 2009–clearly a direct index of the Naga aspiration for peace. That period saw a strategic shift in the attitude of NSCN(K) to themselves engage in peace talk with Government of India by signing a similar ceasefire agreement like NSCN (IM) did in 1997. But before the latter actually came to the peace-talk, it has started talking in divergent tenor, tone, and voice; giving rapacious moments to the rival faction. Most hard-hitting among the Khaplang groups' counter-tremours are the statement at Khehoi (Sunday: 21st March 2010) of its Ato Kilonser, Kitovi Zhimoni that: "possibility of resumption of more bloodshed among the Nagas exist, if a solution less than sovereignty is reached between Government of India and other Naga group." This has made a non-aligned liberal Naga comment: "It seems GPRN/NSCN who were earlier having soft approach for solution without sovereignty and Greater Nagalim have become more of a hurdle than the facilitators."[31]

The Naga public seems to be pining for peace but itself remains pitiably unaligned–if not sharply divided between the two factions. Because, Kitovi Zhimoni earlier declared also at Khehoi on the occasion of the 30th Republic Day of GPRN/NSCN that "If the public gives support for solution to only one group, they will also be responsible for any bloodshed". So as to allay any misgivings among doubting Toms, he quoted past events when the 16-point agreement resulted in distrust and bloodbath among the Nagas. His add-on establishes a cause-effect relationship between 'agreement' as such and bloodshed: "Any accord is signed with India, bloodshed in the State has followed. Do we need another accord to start another round of bloodshed?" Bluntly put, according to him the peace talk has thrown the soft-liner Nagas (unaligned to either faction) in "utter confusion" about "peace" as its possible outcome. A kind of duststorm has been kicked up. This, despite the FNR, which has been, supposedly playing a strategic role to wards reconciliation between the two factions in the ambit of the overall Naga peace.

A final settlement to the Naga issue begs two basic questions, as per informed and elite Naga: the question of Naga sovereignty; and demand for a greater Nagalim. Both (are articulated to) be tied to Naga leadership's larger concerns about their community's political and cultural rights. And the emerging view is that both are self-serving, though in different ways: for Muivah strictly personal or inclusivity of his native Tangkhul-inhabited Ukhrul areas; and for Khaplang to co-extend his boundary as that of India. Any peace-loving Indian will concur to the following proclivities:

"But should absolute sovereignty from the Indian nation State be seen as a necessary condition for preserving their rights? The experience of independent India reveals that most linguistic and ethnic subnationalities have successfully addressed this question within the framework of the Indian Constitution which provides sufficient guarantees for the rights of cultural, linguistic, ethnic and religious minorities. The NSCN should be more flexible on sovereignty issue and negotiate for political autonomy. The federal principle enshrined in the Indian Constitution allows a great deal of flexibility on this front. Similarly, NSCN & GPRN/NSCN must revisit their demands for a greater Nagalim. The region is like a 'salad plate' of different ethnic and linguistic nationalities. Any redrawing of boundaries will definitely trigger unrest in the neighbourhood including in Myanmar where many Naga tribes live."

"People of the State are, as it is, troubled and economically harassed by taxes imposed on everything by the entire UG groups in the State. Another fear cropping up is also inter-factional clashes and killings which have been relatively gone zero in the year 2009 and State has seen the peace."

"An end to insurgency will have enormous impact on the northeastern societies. The Naga rebellion has inspired and influenced numerous insurgencies in the region. Thousands of lives have been lost as a result. Peace in Nagaland will trigger development and trade in the region, unlocking the potentials of its people. We must end the cycle of blood-letting and corruption which has crippled the region's prospects. That will also seal Muivah's and other Naga leaders' place in history provided Government and Naga rebels work out a deal".[32]

16

ONGOING CONFLICT RESOLUTION

In response to the Government of India's policy militant outfits like the United Peoples' Democratic Solidarity (UPDS); Dima Halam Daogah (DHD-Nunisa); National Democratic Front of Bodoland (Pro-talk faction) or NDFB (PT), had entered into Suspension of Operation (SoO) arrangements. Further on, DHD (Joel group), and Karbi Longri NC Hills Liberation Front (KLNLF) have come forward for dialogue to seek resolution of their grievances within the framework of the Indian Constitution, have laid down their arms so as to create a conducive environment for talks.

One will apparently become bewildered by the large-scale "mitosis" in militancy–2-fold (NSCN, NNC, KNF, HPC, DHD, KNF, PULF); 3-fold (KYKL); 6-fold (KCP)–almost on par with the actual mitosis of tribal denominations in their desperate search for identity. If the culturally eventful "mitosis-for-identity" came in the wake of greater awareness for cultural effervescence and benefits for education and employment, the "splintered militancy" also came in apparently for separate rehabilitation benefits (and existentialism). So much so, the mutationary trendy is seen transcending. *Prima facie,* one may concur that this behaviour unravels their superficial character, in sofar as they can come out easily unstuck from the "ship's bottom", without the self-effacing character of a "barnacle". Because by a process of reductionism, it is inferrable that the "others" are serious militants–bent upon fighting till its logical end (sovereignty).

The Central government–itself engaged in working out a modality for "tripartite talk" with the *Lallumba* faction of the proscribed KCP outfit, as also with the numerous Kuki armed groups which have already signed Suspension of Operation (SoO) agreement with the Centre and the State Governments–realized to its dismay that the government is about to face a "pandora's box". On hindsight, one may surmise that the dialogue process will also go on and on, if the snail's progress (1997-2010) of continuous 70-odd Indo-NSCN (IM) talks is any index, with the aging insurgent self-denying a glorious political empowerment. Hence a spokesman of the MHA has announced a GOI decision "not to enter into dialogue with splinter groups or any other new outfits or accept their laying down arms in a public function". In the process, it seeks to deny the public glare and halo-effect being obtained as windfall, undeservingly. Particularly because of the redundance (bordering on malignancy through gross misuse) of their post-surrender status, it became compulsively necessary to stem (fake) insurgency. At least, long-struggling militant groups have to be discerned from year-old groups (out-and-out opportunistic). And what needed uprooting would be the former, rather than the mutated (derelict mitosis) version, which can of course "surrender" as per and become entitled to cash benefits etc under the stipulated terms of the "Surrender-cum-Rehabilitation Policy".

Further, it has clarified that any break-away subgroup of the SoO-signing outfit, will not be entitled to any immunity from CI Ops from the security forces extended to the main outfit. The MHA spokesperson adds: "It has been observed that some of the renegade cadres of the outfits who did not give up their arms have not joined the ongoing peace process and some of them have reportedly formed splinter groups. Security forces shall take concerted action against such splinter groups".

17

EPILOGUE

The foregoing has stressed the ultimate need to cope with the changing values over time and the evolving needs as driven by utter desperation of unfulfilled human needs; prompting some 'peace constituency' protagonists to redefine 'conflict' as an essentially creative element, as explicit in case of it regenerating 'change' either non-coercively (or nonviolently) or otherwise. Naturally, 'militancy', 'insurgency'; etc are all the extremely de-structive (and all-offensive) means, like war. But in the other extreme are those slow and gradual change visualized by the peaceniks. Either way, it embodies the means to change: better still a 'driver to change' (human) social values of welfare, security, justice and opportunities for personal development. John W. Burton goes a step further and considers (non-violent conflict) as a crucial catalyst for social change. But the aim is to prevent and eliminate the violent and destructive manifestations of conflict: [33]

> " . . . [c]onflict, like sex, is an essential creative element in human relationships. It is the means to change, the means by which our social values of welfare, security, justice, and opportunities for personal development can be achieved . . . The existence of a flow of conflict is the only guarantee that aspirations of society will be attained. Indeed, conflict, like sex, is to be enjoyed."

Quite interestingly, the present-day Naxalites of central Hindi heartland are being redefined as 'necessitous man'–following pent-up and long-unfulfilled desires (human needs) and the violence explained away as "From Victimhood to Violence", but certainly not as insurgency or inspired by Marxist-ideology. But such logic also goes to explain the same insurgency in NE India, including Nagaland/Manipur, with of course a tongue-in-cheek overtone (tilted by identity brigands) to ethnicity issues. So much so, that today each identity leader has largely insensitive to the need for harmonious living and the transcending cause of positive (non-coercive) peace, as he has since justified it as an act to protect (defend) the need (once obtained or accessed). But in the process he has obtained (earlier tight but increasingly comfortable) political space so as to empower 'self' and after him others in the same ethnicity. Next, operates the law of 'larger than life' and consequential rivalry among 'other' groups, who seek similar identity brigand. And the cycle goes on; until a merry-go-round comes up all over the region, more as a caravan of groups by others, although they are themselves enjoying the innuendo in a gala carnival environment. As long as this double standard (perverted) innuendo and ultimate slide-back to self aggrandizement becomes 'dominant human behaviour unwillingness to give 'peace' a chance remains true, there will be no forward movement towards greater harmony and sustainable peace. The faction leaders are still more reluctant to smother the fire of distrust. But if the public generally become convinced of peace as the need of the hour, they can very well inject this 'oxygen' to revive the 'political will' of the leadership.

Two things become most foundationally required. Peace among ethnic groups together with stability in inter-ethnic relations are most essential. Having remained in a situation of 'trust deficit' all these decades, it becomes critical for all stakeholders to work for sustainable peace between each and every ethnic group. Civil societies in each ethnic group, by and large, support the goal of peace among ethnic groups. It is important that the trust deficit and the burden of history not be allowed to impinge on the task of moving inter-group relations forward on the road for development.

Trust can be built up through multiple uninterruptible dialogues (and cultural exchange) between civil society organizations, besides taking positive and incremental steps, confidence (trust)-building measures, and most critically through acts of statesmanship by the leaders of each group. A grand reconciliation can only be ensured, in the long run, through engagement at every level: civil society meetings, official dialogues, engagement of political leaders, cooperation between business communities (traders), visits of artists, sportsmen, media, talks between ethnic group leaders on both sides of the divide; and if possible through track II and III engagements at the level of the elite and common people respectively. What is important for the elite is to even set the 'agenda for peace', including associated questions of wherewithal, when, how, etc.

Track II dialogues are designed to move beyond officially stated positions (what can not be morally done) so that a way forward can be found, which will offer alternative approaches to government to come over and render all the needful assistance. These dialogues go a long way to clear any vexed (grey) area between the cognitive civil space of each group. In fact, it provides a road-map for the top leadership to propose new initiatives for peace, amity, accord and sustainable relationship. Effective dialogues could also provide the much-needed oxygen to keep the pro-peace constituency. For that matter, political will of the ethnic leadership rest squarely on the amount of amity, and positive peace content in the public space. Hence Track II and III channels would very much help in bring the cold war time-warp to an end. Once they start interacting, then they will no longer feel, for example, that the dominant group is marginalizing the minority. In spite of the stand-off, traders, artists, and private channels are still interacting, and they will have to continue doing so, having done so since time immemorial during their existentialistic lifeways.

A stand-off also allows plenty of scope for misunderstanding between the identity factions. In fact, the media can never be silenced, and they will float a possible (virtual) 'media war', which adds to the trust deficit, by making the atmosphere more opaque. Hence' a direct reaction/interaction/dialogue would avoid all these misconceptions. In democracy, the much aroused awareness of the natives in a land-locked polity like Manipur will have to run through the run-of-the mill (maze) during transformation from tradition to modernity, still very much behind the mainstream who are now morphing from modernity to post-modernity. Dogmas have to be overcome.

Times have also changed. Humanity will henceforth be no longer as homogenous and symmetrical as they used to be. So much of desperation, poverty, and (demand-supply) mismatch is no longer responded to by a supplier (of labour inputs) in terms of the universal upward-sloping supply-curve, but very symbolically in terms of a "backward-bending supply curve", which means as more are still more are offered by way of wage, labour shrinks. Under such extremely unpredictable times, some amount of violence (terror) will prevail, more or less conforming to what strategists presume, "terror or 'low intensity conflict" will always remain where Home Minister P Chidambaram has established a permanent Homeland Security system. Even the Union Government agency, the Defence Research and Development Organization (DRDO) is now developing systems to meet the challenges of terrorism and low-intensity conflict, now emerging as the most potent threat to security. Being innovative scientific applications in the latest area like nano-technology, Western nations are found unwilling to share in these technologies with India. They range from the futuristic high-energy weapons system and complete network–as nuclear warfare systems, including unmanned aerial vehicles and surveillance systems for penetrating thick foliage. Also being considered are developments over the existing' systems like soldier-centric defence, and yet others as nuclear, biological and coastal surveillance, better communication, terror-proof armoury, jammers and more effective capability in terror attack.

Yes, temporary setbacks have been seen in inter-group accord, amity and trust. But these shortfalls need not be allowed to impinge on people-to-people cooperation in their life-ways. Attempts should also be made to create an atmosphere of free accord, and amity for important stakeholders including academics, journalists, business-men, students, artists and retired officials and elite sections of each group.

The media are playing a critical role in shaping popular perceptions. They have thus a great role to play to help strengthen the 'constituency for peace'. A continuing dialogue between journalists, editors and proprietors of media houses is most needed for that purpose. Throughout the region insurgency seems declining in terms of lesser violence. In Assam, the government cites that the overall law and order situation in Assam has improved in recent months, yet the police and security forces should not be complacent and they should not lower their guard as according to government records, at least seven militant groups are still active in the State and possibility of subversive activities cannot be ruled out. The government announced in the State Assembly that militant groups like the United Liberation Front of Assam (ULFA), anti-talk faction of NDFB, All Adibasi National Liberation Army, Kuki Revolutionary Army, Hmar People's Conference (Democratic), Muslim United Liberation Tigers of Assam (MULTA) and Harkarul-Mujaheedin (HuM) are still active in the state, which is definitely a matter of concern and the security forces should keep a close watch on the situation. Among the active militant groups, ULFA suffered serious setbacks in recent months but the outfit still has more than 800 members. Among the other active groups in Assam, the MULTA and the HuM may pose a serious security threat as the possibility of those groups receiving direct or indirect assistance from anti-India forces cannot be ruled out. The government must take all possible measures to seal the international border to prevent elements of anti-India forces from sneaking into India from Bangladesh.

The latest polarization of views on the state of militancy in public space is that of the mainstream press. Even among these is the headline view that terms it as a business proposition, where both the state and non-state actors are "stakeholders" and "contribute to each other's sinister agenda." So, it alludes, [34]

> "Why should one even imagine of its solution? Will not a solution bring down the entire edifice of the industry? But the reality of all realities is that such insurgency cannot be solved because those who matter do not want a solution. The pretence on the part of both state and non-state actors,

> however, must continue so that they can hoodwink an already looted an tormented people; the militant says he is staging a 'revolution' for the masses whose cause he thinks his outfit alone must champion, while the powerful state actors say they are seriously engaged in the business of 'conflict resolution'. Given the fact of life, for the militant leader the talk of sovereignty, self-rule et al is a mere charade to negotiate the sustenance of his non-state role so as to have the best of both worlds–sympathy from the masses of whom he has fashioned himself as their leader and the undeterred flourish of his stake in the industry of terrorism. It is also a burden because he is always constrained to prove what he is not and what he cannot now be–a leader of the masses. But the advantage is that the politician involved in the industry of terrorism has his own vested interests to further and needs the militant leader while at the same time he must pretend that the militant is an enemy of the state who must be fought and decimated. Both the militant and the politician involved in the industry of terrorism thus expediently exist for the sake of each other."

The Central security forces and the Central government also seems to be converging to the above view. The only regret is that, it took 3/4 decade-long militancy to make the Delhi-NE India gulf. And this, despite multi-layered many-channel administrative connection between Delhi and NE India.

Even if a system of as many as roughly four dozen ceasefire agreements are posited the post-conflict situation looks no less haunted. For the greatest problem will then arise as to who are the first settlers, original inhabitants or autochthones, as opposed to immigrant population or diaspora (floating) population. In fact, it begs the question as to who is autochthone, and who is alien or immigrant after some generations of inter-ethnic marriages as say between autochthones and immigrants. And the problem will be stupendous, if not impossible for Manipur's two-millennia old history-supported population of Manipur (valley); unless DNA or any other scientific tests can be used for the mammoth task. Thus even ceasefire will haul the so-called 'bad peace' a form of negative peace, wherein peace is maintained (make done), neither lasting nor sustainable. On assessment of the plausible post-conflict situation with a system of cease fire operations in place (with as many four dozen UG outfits) in Manipur alone–fortunately none of them left-oriented–even the post-conflict situation will not be problem-solving and, if at all, delay-prone or protracted in implementation. As to the CR progress, the bottomline is that: "just as separatism is a mere slogan *sans* substance under the present circumstances, conflict resolution too is a mere theme for seminars and conferences. The 'conflict' will be resolved only when the infrastructure of the industry of terrorism in the name of insurgency is, dismantled completely. It is here that political will is lacking, which seems natural given the kind of industry in place. Only the State government seems to remain adroitly 'silent' if any they seem to survive, more on the throes of existentialism. "The politician involved in the industry of terrorism has his own vested interests to further and needs tend that the militant while at the same time he must pretend that the militant is the enemy of the state who must be fought and decimated. Both the militant and the politician, involved in the industry of terrorism thus expediently exist for the sake of each other."

What is important for inquisitive analysts would be the "swing factor". Should the Army perceive a swing to lie to the extent to which UG-elements' propensity to lay down arms is imminent from the Suspension-of-Operation agreements signed by some elements with the Government of India, also apparently at the instance of the Army? What is also significant is the essentiality of con sensuality to any such inter-dimensional 'shift' from violence to nonviolence. At heart is the need that one and all have to accede to the new shift. May be, some elements are only paying 'lip-service' to the peace effort, or are they convinced about the 'turning point.' Per-haps, all the 'other' UG elements –belonging to diversified ethnic groups–are after all conjoined twins having similar strategic interests; and in addition, similarly–rather than differently–India-centric.

Given a vertical forum through a vector (as in vector-induced gene modification in BT brinjal), this initiative could be made good use, provided ample (or 360° transparency is assured). Also given that the peace-talk process is rather very long and arduous as seen in Ireland, Spain, Yugoslavia, Palestine, not to mention the Indo-Naga peace talks still hanging in the balance, it becomes necessary to doubly vouchsafe absence of any lurking risk, opaqueness, conviction before even a road-map to the CR process is contemplated.

With the endgame in sight (if not in process) one thing seems certain. Indian Army is the only institution which can be relied for its grassroots involvement for a pretty long time (spanning more or less the last four or five decades) and therefore with what may be termed 'strategic depth' of the problem. And if at all the Indian Army wishes to cash in the changed (conflict) perception, naturally it need reach out to not only the peace constituency in the State but more particularly for cooperation from peace activists for widest possible dissemination of the peace awareness (for the fringe population) and peace process among UGs. After all, the Army can only help through during the initial teething period, for ultimately the people need rise to the occasion.

In a sharp deviation from past practice, the Army needs be lauded on yet another fundamental point, for trying to both prick the public space and also relying and directly interacting with the public leaders, rather than go the 'official' route and rely wholesale on the 'peace-spoilers'. For much beyond ordinary laymen's perception the official constituencies are seen having a vested interest in prolonging the conflict: be it police or official and even government leaders, who always find a lame excuse to blame insurgents while whitewashing own corruption. Else, even Hindi heartland has found Task Force/Special Investigation Team has been caught in camera receiving extortion money. On that dimension, it can be safely argued that even the poor-governance suits their interest. For, only with the help of close-circuit TV these surreptitious crimes could be detected there; certainly that risk-factor is missing. In fact, galore are the instances here wherein Police were found resorting all these nefarious activities, which could not be brought to book for want of evidence (caught-by-camera) and/or risk of offending an armed policemen who could exterminate a public. These 'peace-spoilers' are found among government leaders who have to make 'politically correct statements' often hushing up (short of covering) even acts of sabotage as calculable risk of political game.

One cannot help at this stage reckon the cost of conflict borne by the State, and the missed development opportunities during the last 3 decades or more. It will take much more to level itself on par with other equally placed (neighbouring) states n the Indian Union. But on final assessment, the CI Ops need not have lasted so long a period of over three decades. In a way, it was going astray more like Arreguin-Toft's well-convincingly enunciated "asymmetric war", where "the strong do sometimes lose". Security forces need not only displace insurgents but pave the way for a civilian administration to restore ongoing development programmes (with added gusto and spirit).[35] Even after restoring confidence in the hitherto violence-affected and rights-deprived community in the so-called 'liberated zone' through a broad-based and meticulously-negotiated dialogue process along structured and 'peace-tracks' the security forces need remain proactive in the areas of capacity-building (in league with the civilian administration) through viable farm-cum-agroprocessing and micro-enterprises so that the rehabilitated insurgent personnel finds his own *'locus standi'* in the economy, with moderate assistance and State support. Of course the Ministry of Home Affairs have now learnt it the hard way from the American strategists (and Security Adviser). On hindsight, it holds out so very true that even in anti-Naxalite operations in Chhattisgarh (Dantewada hillock massacre for instance, the security forces have not learnt the underpinning strategy of CI Ops: to "clear, hold and build", howsoever much they might win battles. Only belatedly, in Manipur they seem to realize the need for a parallel D-system of Dialogue, Development and Deterrence. And last but not the least, the entire CI Operations have betrayed a hit-and-miss portrayal, which has aggravated the conflict situation perception from the human rights angle. After all, these outfits do not count as a counterweight against the multitudinous security forces, with advantage of sophisticated weapons and information network. In the absence of any strategists, the security forces could not hitherto be expected to rely heavily for its success on patience, approaching the legendary, *"Fester lente"* [Make haste, but slowly!]– the motto guiding the Roman emperor Gaius Julius Caesar Augustus in his phenomenal success story. Better late than never!

REFERENCES

1. This is amply demonstrated in *The Other Side –Journal of Socialist Thought and Action:* (as quoted in Business Standard March 9: 2010);
2. Gangmumei Kamei: Ethnicity and Social Change.
3. Conflict has an ontological (philosophical) base in human needs, the denial of which causes conflict. Hence the 'objective source of conflict lies in needs not in interests.' Peace research like conflict resolution involves a critique of 'realism' in international relations. In realism, states are in conflict over the proportion of resources they control and, therefore, the power they weild in the international community. Likewise, individuals are also in conflict because individual 'realism', after Thomas Hobbes, is in such a state of nature that life for 'man' would be nasty, brutish and short, and that it would be characterized by a continual war of all against all. (Of the Natural Condition of Mankind; as Concerning Their Felicity, and Misery, *Classics of International Relations,* 2nd Ed. John Vasques; Princeton: 1990: pp213-15) As quoted by A B Fetherston: *From Conflict Resolution to Transformative Peace-building: Reflections from Croatia*: University of Bradford: 2000)

   This picture of an anarchical world was then applied to international relations where in the absence of some check on the nastier side of human nature (a 'check' which states supplied, at the behest of man, within their domestic realms), anarchy had free reign. Cooperation was possible only to the extent that it maximized self-interested goals of individual states and minimized risk to sovereignty. Violent conflict in the international arena can only be contained and controlled.(Kenneth Waltz: *Man, the State and War:* New York: 1959; and D A Baldwin (Ed): *Neorealism and Neoliberalism: The Contemporary Debate: New York:* 1993)
4. *Ahong-Achao* (or abundance amidst prosperity) used to be the generic Manipur monarchy's pledged catch-slogan. In that land characterized by abundance and all round prosperity, people led a well contented life with utmost cordiality and cooperative (helpful) attitude to 'others' mutual reciprocity and trust. Even thuggery or any petty use of force (violence or violation of the monarchical power) almost ruled out among the then Manipuris out-and-out obedient and god-fearing (authority-reverent). This 'positive peace' transcended almost till the Second World War, when Manipur saw modern warfare.

   As of then, every farmer had sufficient farm land whose output more than filled his granary, with large residential complexes (more or less enclaves) with extensive kitchen-garden growing vegetables all through the year and with large fish-ponds. So much so that even in the marketplace either rice, fish, or handicrafts, and other merchandise used to be sold very cheap. The public (For more see H Dwijasekhar Sharma *et al:* Economic History of Manipur During British Raj. In H Dwijasekhar Sharma (Ed): *New Insights into the Glorious Heritage of Manipur:* Akansha Publishing House: New Delhi: 2009.)

5. Even otherwise, as is very well known, the government is pledged to bring about 'social change" expeditiously so that modernity could be engendered in the polity, wherein government is confronted by various vested elements solely arguing on the basis of preserving the 'tradition'. This is best summed up as either: "Modernity is the scalpel which will severe (all surviving) traditions before the polity attains and prepares itself to achieve 'Post-Modernity'." This underlying idea was spotlighted by Prime Minister Dr. Manmohan Singh on 17th-January, 2010 while inaugurating the 150th anniversary of St. Xavier's School in Kolkata, when he called for stress on "rational thinking and scientific temperament to counter dogmas."
6. *Also see Endnote 5: Intractable Loktak is yet another Achilles' Heel*, at pp.88-90.
7. In fact, that doubt (overlapping) still lingers on. Nearer home, Gambhir Singh is testified to have "subdued" various savage tribals in and around Kohima, leaving behind according to British accounts and other sources even his 'Footprints' (since demolished during advent of Naga insurgency in post-Independence period). Thus it can only mean 'suzerain' what China is now acknowledged to have over Tibet by responsible British authorities (themselves having been involved in the then demarcation of Indo-Chinese border and thereby of the position of Tibet) that China.
8. For more *see Sec 9*: Language as a Means of Unifying Tribal Identity Seekers: T S Gangte: *Kuki-Naga Ethnicity in Current Manipur Poility*. In H Dwijasekhar Sharma (ed.) Ibid;
9. BBC Interview *Ibid* p.14.
10. *The Hindu*: Editorial: 3rd Aug 2005.
11. In her Hudson Memorial lecture: "Cyclic Time and Linear Time in Ancient India" (organized by Prakriti Foundation) the eminent historian Romila Thapar has differentiated between 'cyclic time' and 'linear time' or "arrow of time" and stated that ancient India had both these two concepts. She argued that cyclic time prevented the evolution of a sense of history. There is a close link (association) between linear time and (the sense of) history, though the two concepts may intersect very interestingly on some occasions. Linear time is the functional concept, "dependent on human activity". Based on that linear concept of time she would add that different segments of society may view their past in different ways. "If time were to be seen as a metaphor of history, many dimensions in historical intersections required to be explored." (*The Hindu*: January 8th, 2010: *p12)*. However any identity ideologue (brigand) will now demand that they talk of history as events unfold in a linear fashion, and therefore that in a basic sense progress is inevitable.

    But the reality is much too convoluted to conform to such simplistic models. One hardly reads a newspaper the way one reads poetry (or serious books on academic), although it may be reversible (poetry being read as newspaper) in some rare instances. The real world is a cluster of probabilities and alternate pathways–all outcome of creativity as diverse as poets' genius and the natural law governing poetry. (For more also see section 11: Shadow of Terrorism.) Most historians (and now identity brigands of an imagined 'Naga' community state what looms large, gathering the history of sensibilities transcending in folklore and legend. But they need be warned of the possible fallacy created by the "arrow of time" or perhaps inherent in its concept. Rather they read a newspaper, as if poetry.

The 'inclusive spirit' and the 'open mind' that is free of xenophobia and cultural prejudice might have fuelled some poets or art-forms. But in the last resort, there was no renaissance which could enrich and resonate the Manipuri mind. Devoid of this essential phase in the intellectual history of a group naturally the group would have much more 'ironies and coincidences" to underpin a resurgent renaissance. As a result, the trajectory of a nationhood based on 'imagined history' (more or less the missing span(s) in the trajectory of the 'arrow of time' has to traverse like that of a poet. Just outside the Kangla, it took three decades since the Statehood day (21st January 1972) to form a sustainable government for Manipur's politicians, after their century long "historic indecision".

12. Paper contributor's personal collection.
13. "Supremacy" is the term chosen here, because in the days of yore the clans or ethnic groups were certainly "un-formed" polities, when it will be hard to conclude Gump the gun) that any horde of invader would establish permanent supremacy or a still chaste term "sovereignty". At the most any conqueror will be treated more as a "feudal overlord." Oxford dictionary interprets suzerain as "1. a feudal overlord; 2. a sovereign state". Thus, it can only mean 'suzerain' what China is now acknowledged to have over Tibet by responsible British authorities (themselves having been involved in the then demarcation of Indo-Chinese border and thereby of the position of Tibet) that China.
14. W Damodar Singh, IPS (retd) Paper: *Manipur - A Nation:*
15 *Pawsey Papers:* Cambridge South Asian Archives, Cambridge University, Clow to Pawsey: 26th April 1947, Box No.1).

    [In fact, reminiscent of the (primitive) tribals of Andaman and Nicobar Islands, who remain till date more or less as living 'museum pieces', because any exposure with the non-tribal world of even making a road for outsider to pass through their preserve has led to their extinction (through infection etc.). God saved Manipur and North East India from such gross care.]
16. For more detail see paper entitled *'Manipuri Nata Sankirtan:* Endnote 18: First Kirtan at Lamangdong Vishnu Temple: Vol I of this Compendium
17. C J Lyall's conviction, since put down as Introduction in T C Hodson's *The Meitheis*: p xvii;
18. China's 'globalization outlook'.
19. Ascribed to Vice-Admiral Arun Prakash PVSM, AVSM, vr C, VSM.
20. See G P Singh: *Religious and Cultural Syncretization of the 18th Century Manipur*: in Volume 1 of this anthology;

21. R S Jassal, the present paper contributor has authored various articles on Peace-People and the Periphery incorporating similar views in the North-East Herald (Kohima), *The Sangai Express*, The Imphal Free Press (Imphal)
22. Gangumei Kamei while addressing at the Centre of South Asia, JNU, New Delhi;
23. For more on the Verelst-Gourashyam Treaty and unsuccessful mission of the British effort due to the "Historical significance of the 'natural borders' of Manipur" and contents of the Treaty see H. Dwijasekhar Sharma (ed): *Ibid.* Chapter 6: Endnotes 3 and 7).
24. Also see another paper: *The Advent & Culture of Manipuri Muslims* by Abdur Rahman in Vol II of this anthology.
25. For more on poverty, deprivation, caste inequity, etc. see Vol II of this anthology: paper entitled *The Advent & Culture of Manipuri Muslims,* Appendices-I: The Muslim Deprivation; II-Economics of caste inequity, III-Tendulkar Poverty Line; IV- Differential reach of primary education.
26. King Khagemba (1652-56).
27. Major Maxwell, the first British Political Agent ran the entire Manipur administration directly between 1891-1907, during the infancy of Raja Churachand Singh. Even thereafter, when the king ruled in his name, he did not have any overriding power over the Manipur Hills - whose administration was vested to the British, holding the post of President of the Manipur State Darbar. In their bid to keep the tribal areas un-blemished by any element of change through their colonial administration, they also banned on grounds of budgetary constraint the erstwhile holding the traditional Mera Haochongba (Valley-Hill Meet) event at the palace grounds which led to generating amity and fraternity among hill and plains peoples. When they modernized Manipur and introduced a budget by introducing cadastral survey ofland and imposition of land revenue, they further stopped short of extending it to Jhummia land in the hills. *(Also see A Kamson: The Mera Haochongba Festival: The Traditional Hill-Valley Interface — The Carnival of Manipur* in H Dwijasekhar Sharma (ed) *Ibid).*
28. C J Lyall: *Introduction to T C Hodson's The Meitheis*: p xv.
29. Lyall: *Ibid*: p vii.
30. Satchidananda: *Indian Poetry: Modernism and After*, SahityaAkademi: p.2.
31. *The Sangai Express* English Edn: 6 April 2010: p2.
32 Excerpted from Robert Lanky, Go for peace, *The Sangai Express*, English Edn. 6 April, 2010.
33. John W. Burton: *World Society:* Macmillan, London, 1972.
34: *The Assam Tribune*, April 1, 2010.
35. ADNI: The True Hue of NE Insurgency, *The Sangai Express*, English Edn: 3 April, 2010.

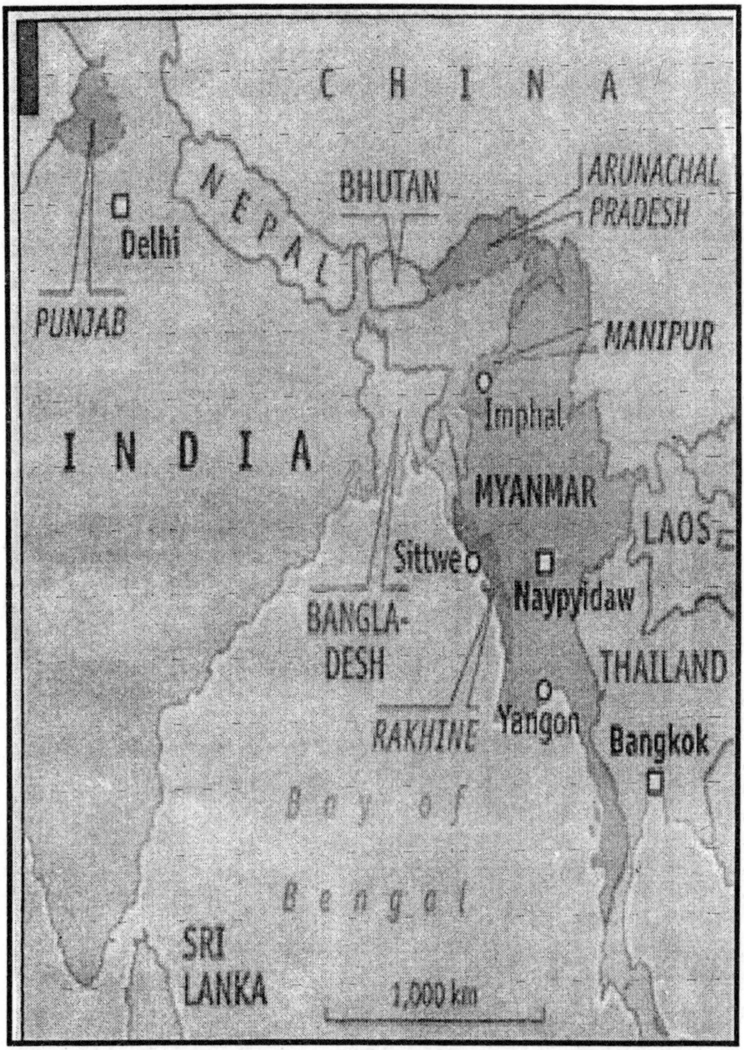
CHINA
NEPAL
BHUTAN
ARUNACHAL PRADESH
Delhi
PUNJAB
MANIPUR
Imphal
INDIA
MYANMAR
LAOS
Sittwe
Naypyidaw
BANGLA-DESH
THAILAND
Yangon
RAKHINE
Bangkok
Bay of Bengal
SRI LANKA
1,000 km

Table 1

I. INTRINSIC CONFLICT

State-wise list of Outfits with last known strength and No. of weapons

(a) MANIPUR

Sr No.	*Name of Outfit*	*Strength/ No. of Weapons*	*Comfined to*
(1)	*(2)*	*(3)*	*(4)*
1.	PLA-RPF	1,500 and 700	Imphal, Bishnupur and Thoubal (Sept 1978)
2.	UNLF (now MPLF)	1,000 and 600	Imphal, Bishnupur and Chandel (Nov 1964)
3.	PREPAK (Red Army)	200 and 150	Imphal, Bishnupur & parts of Chandel (Oct 1977)
4.	KCP	100 and 50	Thoubal and Imphal districts
5.	KCP (Noyon)	370 in all (items 5–9)	
6.	KCP (Lamphel)		
7.	KCP (Lallhumba)		
8.	KCP (MC)		
9.	KCP (VC)		
10.	KCP (L. Khuman)		
11.	KCP (MTF)		
12.	KCP (Nogdren Khomba)		
13.	KCP (MC. Tongaba)		
14.	KCP (Maoist)		
15.	KYKL (0)	170 and 100 weapons	Bishnupur/parts of Imphal district
16.	K YKL (T)	50 and 20 weapons	Imphal and Thoubal district
17.	KYKL (MDF)		
18.	UKLF	100 and 70 weapons	Chandel, Moreh, Tengnoupal
19.	KNAIKNO	200 and 90 weapons	Parts of Chandel, Churachandpur & some influence in Sadar Hills
20.	KNF (MC)	250 and 150	Senapati,Churachandpur

(Contd...)

(1)	(2)	(3)	(4)
		weapons	(with ZRA) and Sadar Hills
21.	KNF (PC)	20 and 15 weapons	NC Hills (Assam) adjoining Jiribam, and Sadar Hills
22.	KNF (S)		
23.	ZRAIUPF	400 weapons	Churachandpur district
24.	HPC (D)	60 and 20 weapons	Vangai range Jiribam though surrendered to Assam Government in 1993 but re-emerged.
25.	HRF	40 and 20 weapons	Churachandpur
26.	HNA (Hmar National Army)		
27.	DHD (Jewel)	30 and 10 weaPlJns	Vangai range-South Cachar adjoining Jiribam
28.	DAD (Nunisa)	125 and 50 weapons	NC Hills, adjoining Jiribam
29.	NSCN (IM)	800 and 500 weapons	Ukhrul, Senapati and Tamenglong
30.	NSCN (K)	300 and 100 weapons	Ukhrul, Senapati and Tamenglong
31.	NNC (Adino)	substantial No.of sympathizers with negligible No.of weapons	
32.	MNRF	50 and 10 weapons	Ukhrul South
33.	NNW (PSG)		
34.	NNC (Khadao)	100 and 50 weapons	
35.	CKRF-CKLA	Chin Kuki Revolutionarv Front	
36.	NEMF	North East Minority Front	Muslim group
37.	KDF	(KCP-splinter group)	Kangleipak Defense Force
38.	KRA (Kuki Revolutionary Army)	50 and 20 weapons	Sadar Hills
39.	CHIMBY-KCP		China Himalayan (Range) India Bay of Bengal Myanmar-Chimby
40.	PULF (under NSCN- IM protégé	400 and 60 weapons	Thoubal, Dimapur, (having need-based links with all groups)

(Contd...)

(1)	(2)	(3)	(4)
41.	UKRA	30	United Komrem Revolutionary Army
42.	UNPC	18	United Nagas Peoples' Convention
43.	USRA	20	United Socialist/ Revolutionary Armv
44.	UTLA (United Tribals Liberation Armay)		
45.	ZUF (Zeliangrong United Front)		
46.	SPLA (Sinlum Peoples' Liberation Army)		
47.	UPPK (United Peoples' Party of Kangleipak)		
48.	MPLF — Seperates to UNLF Kanglei Yawol kanba Lup		
49.	UKLA (MC)		
50.	MNLF — New Organisation of Manipur National Liberation Front		
51.	NNW (PSG)		
52.	PREPAK (VC)		
53.	PREPAK (GS)		
54.	KCP (Nando)		
55.	KCP (KK Nganba Group)		
56.	DHD (Nunisa)		
57.	NDFB (PT)		
58.	DHD (J)		
59.	UPDS		
60.	KLNF		

(b) NAGALAND

Sl No.	*Name of Outfit*	*Strength/ No. of Weapons*	*Comfined to*
(1)	*(2)*	*(3)*	*(4)*
1.	NSCN (IM)	2200 and 1600 weapons	Sema, Chakesang, Dimapur, Angami, Lotha, and Kiphire areas
2.	NSCN (K)	2000 and 1000 weapons	Mon, Tuensang, Ao, Phom, Sham- mator,Kiphire areas(including northern Myanmar)
3.	NNC (A)	.....	Angami,Chakesang, Lotha & presence in Peren district
4.	NNC (K)	.....	Ao, Sema, Sangtam, Chang and Dimapur areas
5.	KYKL (0)	...	Dimapur, Ghaspani & along NH 39
6.	KNF	.....	Kuki dominated areas of Dimapur and NC Hill including Karbi Anglong
7.	ISI activists	50 (Estmated)	Dimapur, NC Hills backing needy groups

(c) ASSAM

(1)	*(2)*	*(3)*
1.	ULFA	United Liberation Front of Assam
2.	NDFB	National Defence Force Bodo
3.	UPDS	United Peoples' Democratic Solidarity
4.	KLO	Kamtapur Liberation Organization
5.	BLT	Bodo Liberation Tigers
6.	DHD	Dimasa Hoalim Deogah
7.	KNV	Karbi National Volunteers
8.	RNSF	Rabha National Security Force
9.	KRLO	Koch Rajbanshi Liberatin Organisation
10.	KPF	Karbi Peoples' Front
11.	TNRF	Tivia National Revolutionary Force
12.	BCF	Bircha Commando Force
13.	BTF	Bengali Tiger Force
14.	ACF	Adivasi Seurity Force
I5.	ASF	Adivasi Security Force

(Contd...)

(1)	*(2)*	*(3)*
16.	AAASS	All Assam Adivasi Suraksha Samity
17.	GTF	Gorkha Tiger Force
18.	BVYLF	Barak Valley Youth Liberation Front
19.	MULTA	Muslim United Liberation Tigers of Assam
20.	MULFA	Muslim Liberation Front of Assam
21.	MSCA	Muslim Security Council of Assam
22.	ULMA	United Liberation Militia of Assam
23.	ILAA	Islamic Liberation Army of Assam
24.	MVF	Muslim Voluntary Force
25.	MLA	Muslim Liberation Army
26.	MSF	Muslim Security Force
27.	ISS	Islamic Sevak Sangh
28.	IURPI	Islamic United Reformation Protect of India
29.	UMLF A	United Muslim Liberation Force of Assam
30.	RMC	Revolutionary Muslim Commandoes
31.	MTF	Muslim Tiger Force
32.	PULF	Peoples' United Liberation Front
33.	AS	Adam Sena
34.	HuM	Harkul Mujahtdeen
35.	NELF	North East Liberation Front
36.	STF	Samtalil Tiger Force
37.	KPLT	(Karbi Peoples' Liberation Tigers)
38.	SPLA	(Assam aso N.C. Hills)
39.	KNLF	
40.	UPDS	
41.	DNDF	(Dimasa National Democratic Front)

(d) ARUNACHAL PRADESH

(1)	*(2)*
1.	Liberation Tigers of Arunachal
2.	National Liberation Army
3.	United Liberation Tigers
4.	Revolutionary Army of Arunachalis
5,	NSCN (1M) in Tirap
6.	NSCN (K) in Tirap

(e) TRIPURA

(1)	*(2)*
1.	NLTF
2.	ATTF
3.	TPDF — Tripura People's Democratic Front
4.	NLFT — National Liberation Front of Tripura

(f) MIZORAM

(1)	(2)
1.	BNLF (BRU)
2.	Arakans Liberation Party

(g) MEGHALAYA

(1)	(2)
1.	Heyiewtrao National Liberation Force
2.	ANVC (Achik National Volunteer Corps)
3.	I KSO Khasi Students' Organization (actively involved)

Source: Official sources.

N.B. Government of India may consider opening of a separate office/agency in Manipur like SoO Advisory Group on the lines of the CFMG since problems in Manipur are different from those in Nagaland.

Table 1. A (i)

EVOLUTION OF MEITEI UG GROUPS

UNLF (1964)

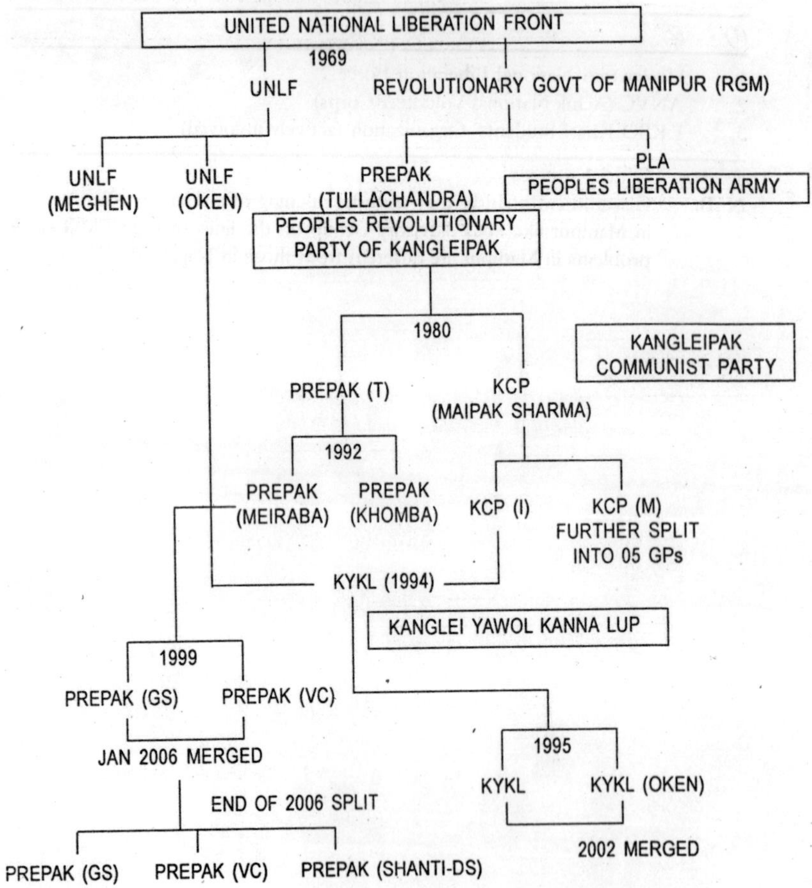

Table 1. A (ii)

PLA FIGHTING WING

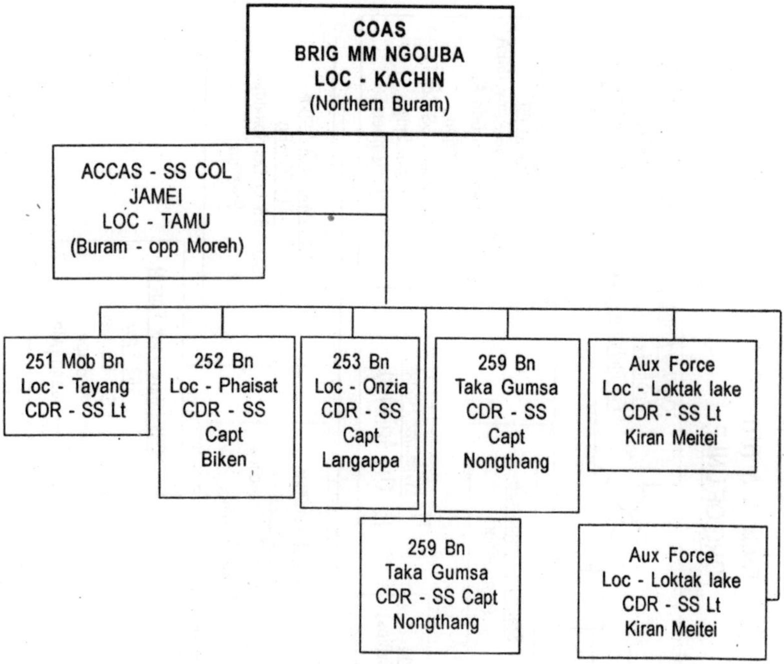

AUX COMD

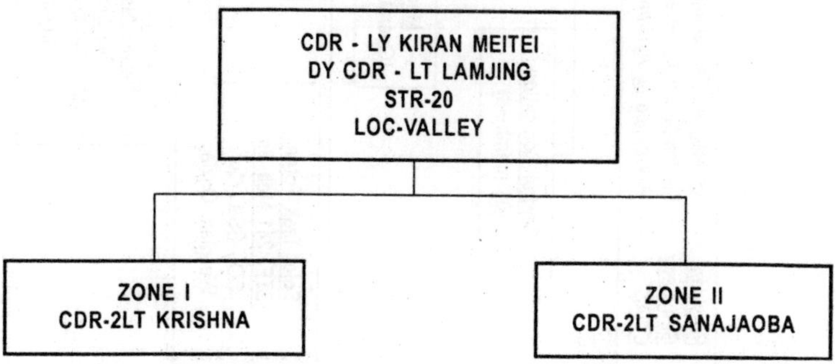

Table 1. A (iii)

ORG OF UNLF

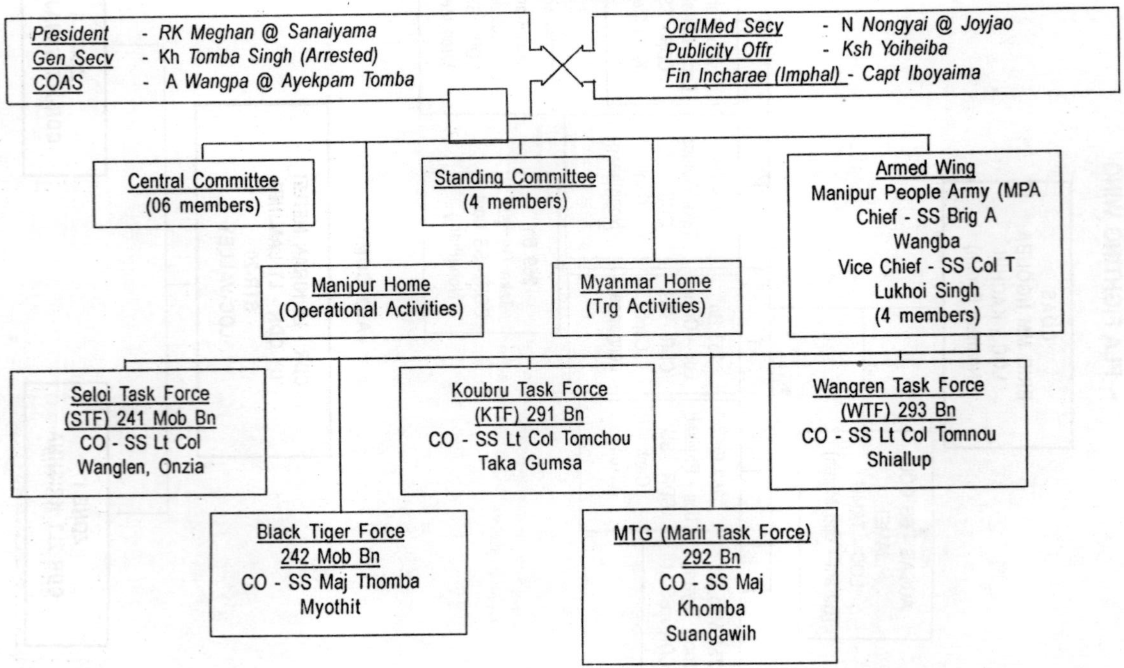

Table 1. A (iv)
Important Days of Militant Outfits

Sl. No.	*Gp.*	*Event*	*Date*
1.	KNA/ KNO	Raising Day	15 Jan
2.	NSCN (IM)	Raising Day (both faction)	31 Jan
3.	MPA	Raising Day	09 Feb
4.	PUlF	Independence Demand Day	25 Feb
5.	RPF	Raising Day	25 Feb
6.	MPLF	Raising Day	01 Mar
7.	NSCN (IM)	Republic Day (both faction)	21 Mar
8.	PLA/ RPF	Martyr's Day	13Apr
9.	KCP	Raising Day	14 Apr
10.	KYKlL	Raising Day	25 Apr
11.	ZRO/ ZRA	Raising Day	07 May
12.	PULF	Raising Day	11 May
13.	KNF	Raising Day	18 May
14.	KNC	Raising Day	10 Jun
15.	KCP	Black Day	01 Jul
16.	NSCN (IM)	Victorious Day	09 Jul
17.	KLA	Raising Day	22 Jul
18.	NSCN (IM)/NNC	Independent Day (both faction)	14 Aug
19.	KYKL	Martyr's Day	18 Aug
20.	UNLF	Oppose Indian Annexation Day	21 Sep - 15 Oct
21.	PLA	Raising Day	25 Sep
22.	NSCN (IM)	Agony Day (both faction)	27 Sep
23.	PREPAK	Raising Day	09 Oct
24.	KNF	Martyr's Day (both faction)	12 Oct
25.	UNLF	National Resolve Day	15 Oct
26.	RREPAK	Remembrance Day	12 Nov
27.	UNLF	Raising Day	24 Nov
28.	KRA	Raising Day	27 Dec

Table 1. B
Kuki Tangle

Kuki identity, refashioned nearer to its historical status, also sets the grounds for tangible deliberations for a stable political future for the people. Besides 'Any Kuki Tribes', an important factor augmenting Kuki unity is 'Suspension of Operations' signed on 10 August 2005 between the Army and KNO, followed by another SoO on 22 August 2008 signed among Government of India, KNO and the state Government of Manipur. Consequent to SoO, twelve armed wing; representing every Kuki clan, now share the KNO political umbrella. They are: (i) Kuki National Army, S Robert Haokip, CAS; (ii) Kuki National Front (Military Council), TH German, C-in-C; (iii) Kuki National Front (Zogam), Joshua Haokip, C-in-C, (iv) United Socialist Revolutionary Army, L Vaiphei, C-in-C; (v) Hmar National Army, LS Lungtrau, C-in-C; (vi) Zomi Revolutionary Front, PS Hangshing, C-in-C; (vii) United Komrem Revolutionary Army, T Karong, C-in-C; (viii) United Minorities Liberation Front, K Khaling, C-in-C; (ix) Zou Defence Volunteer, Pakap Anthony, C-in-C; (x) Kuki Liberation Army, K Wilson Touthang, C-in-C, and (xi) Pakan Reunification Army, Zecky Anal, C-in-C. Kuki unity of this nature is unprecedented. It is the first of its kind in nearly half a century.

The United People's Front, formed in 2006, is another umbrella organisation comprising the same ethnic people. As evident in the organisation's name, 'unity' is the proclaimed motto. UPF's members are (i) Zomi Revolutionary Army, Jackson Paite, C-in-C, (ii) Kuki National Front, T Samuel Haokip, President, (iii) Kuki National Front, Rocco Thangboi Kipgen, President, (iv) United Kuki Liberation Front, SS Haokip, President, (v) Kuki Revolutionary Army, David Hangshing, Chairman, (vi) Hmar People's Convention (D), Ropuia, President, (vii) Kuki Liberation Army, Timothy Khongsai, C-in-C. UPF is dominated in number by groups who profess Kuki as their identity, even though all the groups are ethnically Kukis. Sensitivities to the advocates of the terminology Zomi and sentiment of Hmars appears to be the motive for adopting a neutral term 'United' to represent the organisations.

UPF's political objective is identical to KNO's in terms of the people and land. On the one hand, pre-occupation with an alternative to the terminology Kuki, and on the other hand, considerations of regional issues seems to be an obstacle in the two umbrella organisations shaking hands to forge unity. It has yet to dawn upon the minds of the various leaders that the ideology of land and ethnicity is paramount over nomenclature. No matter what one may identify herself or himself as, at the end of the day 'blood is indeed thicker than water'. It would not be untrue to state that enough fratricidal blood has been shed owing to unbridled priority given to regionalism over nationalism.

Extract is borrowed verbatim from a recent book ***Kuki Society*** *Past, Present, Future,* Published by Kuki Research Forum - 2011.

Table 2

II. Extrinsic Conflict: Myanmarese Outfits:

A. Existing in Ceasefire Agreement with SPDC

List of Outfits with last known strength and no. of weapons

No	*Name*	*Remarks*
1.	BCP	Burma Communist Party
2.	KIO	The Kachin Independent Organisation
3.	Mong Tai Armv	Tai Revolutionary Council
4.	MNDA	The Myanmar National Democratic Alliance
5.	MNSP	The Mvanmar National Solidarity party
6.	NOAA	The National Democratic Alliance Army
7.	SSA	The Shan State Armv
8.	NDA	The New Democratic Armv
9.	KDA	The Kachin Defence Armv
10.	PNO	The Pao National Army
11.	PSLP	The Palung State Liberation Party
12.	KNG	The Kayaw National Guard
13.	KNPLF	The Karenni National People's Liberation Organization
14.	KNLP	The Kayaw New Land Party
15.	SNPLO	The Shan State Nationalities People's Liberation Organization
16.	KNPP	The Karenni National Progressive Party
17.	NMSP	The New Mon State Party

B. Operating Outside any Ceasefire Agreement with SPDC

(a) Major Outfits: List with last known strength and No. of weapons

No.	*Major Group's name*	*Strength*	*Remarks*
1.	Karen National Union	2500-3000	Supported bv US and Thailand
2.	Chin National Progressive Party (CAN)	400-800	Operating in league with NSCN (1M)

(Contd...)

No.	*Major Group's name*	*Strength*	*Remarks*
3.	Karen National Progressive Party (CPB)	500	Aims for an 'Independence Kayah State'
4.	Communist Party of Burma	3000-2000	Covertly kept alive to malign National League for Democracy (NLD)
5.	National Union Party of Arakans (NUPA)	Not known	Its armed wing is known as Arakan Army (AA)
6.	Arakan Liberation Party (ALP)	Not known	Based in Mizoram (suspected)
7.	Shan State Army (SSA)	600	Armed wing is known as SURA

(b) Minor Outfits: List with last known strength and No. of weapons

No.	*Minor Group's name*	*Strength*	*Remarks*
I.	Arakan Rohingiya Islamic	600	Aims for "Independent State of Arkanistan"
2.	Rohingya Solidarity Organization (RSO)	400	Supported by Muslim countries
3.	Rohingya Islamic Patriotic Front (RIPF)	20	Supported by Muslim countries
4.	Itiahadul Mujahidin of Arakan (IMA)	20	Supported by Muslim countries

Source: Official and other authentic sources

ENDNOTES

1. Independence Movement - Naga Hills (NAGALIM):

A. Independence

- Statesman, Monday, August 4, 1947: "NNC demands Independence of Nag as in the Naga Hills district (area: 4,269 sq. miles)";
- 1951, May 16: *Plebiscite in Naga Hills district area* (Forbidding hill districts of Manipur);
- 1954: September 18: Phizo formed Free Government of Nagaland with Hongkhim as President (excluding Manipuris);
- 1955: The Indian Army starts counter-insurgency operations in Naga Hills district, Assam;
- 1957: *As many as 1,760 delegates from Naga Hills & Tuensang formed Naga Peoples' Convention* (excluding anybody from Manipur);
- 1963: A new State of Nagaland was formed;
- 1980: Advent of National Socialist Council of Nagaland (NSCN);
- 1995: June 12: Meeting between Prime Minister of India and the NSCN (IM);
- 2005: Condescension to *autonomy;*

II. Territory

- 1875: The Britishers encompassed some scattered tribes as Naga and formed Naga Hills District of Assam - the first consolidation;
- 1947: Demand for independence of Nag a Hills district (4269 sq. miles);
- 1957: Formation of a greater Naga Hills district with Tuensang from North East Frontier Agency (NEFA), then entrusted to Governor of Assam;
- 1963: Formation of State of Nagaland annexing Dimapur region (hitherto under Dimasas of Cachar);
- 1992-93: *Demand of Greater Nagaland (Nagalim)* encompassing 1,20,000 sq. km (by by claiming to annex territories from Manipur, Assam, Arunachal Pradesh and Burma);
- 1997: Government of India signs Cease-fire agreement at Bangkok respecting Nagalim.

Source: This table. with excerpts in this section) are in-sourced from N. Sanajaoba: *Nagalim-India Federation? Is Sovereignty out of the Agenda?*, *The Poknapham Daily*, 6 Sept 2005) -

2. Indo-NSCN (IM) Peace Talks

"The six-month extension of the ceasefire between the Centre and the National Socialist Council of Nagalim (Isak-Muivah) from July 31, contrasting with the customary one-year extension since the signing of the truce in 1997, speaks of the hurdles that have come in view in the Naga peace process. The intense negotiations held in Delhi between January and June 2005 ended in an impasse. The demand for a shortened truce extension, which came from the NSCN (IM) seems to be an expression of unhappiness over the lack of substantive progress. After more than a dozen rounds of talks between the NSCN (IM) general secretary Thuingaleng Muivah and representatives of the Centre, as well as several discussions between him and Prime Minister Dr. Manmohan Singh, Congress President Mrs. Sonia Gandhi and Home

Minister Shivraj Patil, it is clear the two sides are nowhere close to resolving the main insurgent demands: 'sovereignty' and a 'greater Nagaland' carved out of ethnic Naga areas in neighbouring States. Making a hostage of the eight-year-old truce, which has brought tremendous relief to the people of Nagaland, is not the rational way of arriving at a settlement of the dispute."

"True, the NSCI (IM) no longer speaks about sovereignty. This is not without significance. However the new demand for a 'special federal relationship' outside the framework of the Indian Constitution is virtually a proxy of for the secessionist demand. In the BBC interview, Mr. Muivah made a case for a confederation-like arrangement that would give Nagas control over everything other than defence, foreign affairs, communications and finance (plus a say in the first three areas insofar as Naga interests are affected). On the other impossible demand, Muivah has not budged an inch. Although the NSCN (IM) is not directly involved in the continuing Naga blockade of Manipur over the Ibobi Singh Government's declaration of June 18 as "State Integration Day", it has exploited the protest to underline the urgency of the demand for integrating all Naga-speaking areas in the North-East in one entity. The Naga leadership must realize that the Centre can concede this demand only at the cost of setting off explosive conflicts over territory in Manipur, Assam and Arunachal Pradesh. The silver lining is that just before Mr. Muivah left for Amsterdam the city of his self-exile he put aside his impatience and placed on record his faith in the United Progressive Alliance Government's commitment to a peaceful solution in Nagaland. He also promised to return for more talks. The hope is that the NSCN (IM) leadership, which is scheduled to confer various shades of Naga opinion on the developments in the peace process in Bangkok in the coming weeks, will reiterate its preference for talking over fighting, even if no basis or common ground had been identified for resolving substantive issues."

3. 'The Ghosts of History'

Mutual recriminations in Armenian-Turkish community in Istanbul — as in any other identity-tangled communities elsewhere in the globe–

almost embody "the left-over of the sword" according to the Christian Armenian-born Fethiye Cetin, who was brought up as a Turkish-Muslim. She was just one of the forlorn survivors in the mass-killing of more than one million Armenians by the Ottoman Turks between 1915 and 1918. Cetin herself began to see the world "with different eyes". Now 60, she recites stories of her own mother "burying their own identity - and the horrors they had seen deep inside them." The ghosts of history have since been confronted by the Armenian-Turkish community in Istanbul and the country's leading human rights lawyers believe that "a seminal moment has arrived in which Turkey and Armenia can finally .... overcome one of the world's most enduring and bitter rivalries." Herself one of the seven million strong Armenia diaspora in New York, hence with a global perspective, she continues to be hopeful of "healing generations of acrimony" through soccer. She remains hopeful that even landlocked Armenia's economic isolation would see a silver lining, once Turkey joins the European Union, although Turkey's genocide past remains a crucial issue. On her part, she remains a peacenik. Following her grandmother's wish, "May those days vanish!", she omitted the word 'genocide' from her book because using the word erected a roadblock to reconciliation. (Excerpted from Dan Bilefsky: 'Overcoming mutual recriminations': *The Hindu*, January 11, 2010)

4. 'Porous Border' Staggers Manipur's Insurgency

With no intention to show off any 'commitment lag', the Manipur Director General of Police, Y Joykumar Singh contended in a "Meet the Press" on 17-02-2010 "that activities of UG groups and other insurgency-related problems in Manipur would be reduced by 80 per cent if the neighbouring border of Myanmar is effectively sealed." However, according to him, the envisaged 370-km-long Manipur-Myanmar border fencing is still a long way off; and (only the survey work) is now "under way", but that within 2-3 years "effective border-sealing" would become possible once the critical portion of the border, where the militants cross through, could be fenced off and also a border road be constructed to ensure protection of the border area. Joint operations of the Army and the Assam Rifles in 2005 and thereafter in 2008 (Parliamentary election-eve) have dismantled many of their "permanent or secured camps", as in Loktak lake/National (Sangai) Sanctuary area in Manipur, but their top leaders are safely holed up in their havens across the border. Since these outfits have been operating from their base camps across the border, mostly in Myanmar, and to some extent even in Bangladesh, whose political leadership has not been hitherto so helpful in evicting these outfits, it has added to the woes of the State Police set-up, and for that matter, even of the Central security forces (as also the Union Home Affairs and other involved Ministries). Pithily put, Manipur's top cop's plea is that flushing them out from Manipur is useless, unless Myanmarese and Bangladeshi authorities complement, as driven home by Assam upon Union Government.

An equally, if not more disconcerting, observation has been made casting aspersion on "more bloodshed" because of "the presence of different armed groups and the complex nature of Manipur society's composition". The UG groups have been established more or less on ethnic-wise basis: at least one for an ethnic group. In a way, the much-appreciated 'cultural mosaic' of Manipur–while there was no identity-based conflicts–has become its undoing, now that they are out conflicting with each other. This becomes convincing, once one looks around in Nagaland (and also Senapati and Ukhrul districts of Manipur bordering on Nagaland) where inter-outfit rivalry and annihilation has climaxed, necessitating the civil society to intervene (of course) and help State law enforcement machinery to de-scale down movement of these armed outfits (outside their camps).

A bye-product of insurgency by way of proliferation of various extortion activities has been underscored by way of the large scale collection of illegal taxes now being collected by even those UG groups in Manipur, operating without any ideology, besides those being collected by NSCN (IM) and NSCN (K) from vehicles plying along the Imphal-Dimapur road (NH-39). Admittedly, the deteriorating level of governance (or on political leaders) also hinted, which has also undermined the public confidence level on political leaders, who fell short of public aspirations to deliver. For instance, any area-domination operation is to be followed up by very fast track and ultra-effective civil administration (development) actions, which are not forthcoming. For instance, the permanent evacuation of the Loktak Lake from UG hands is still eluding the State government, thereby they are now requisitioning hovercrafts for regular patrolling. Manipur Police is often found wanting in sensitive counterinsurgency operations, not being equipped with sophisticated equipments as are those in rest of the country.

More interestingly the DGP preempts to speak about the pressure being built upon the UG groups to come out to the negotiating table for talks. Hopefully more groups, than the few Kuki and Meitei groups, could come forward. And on post-conflict measures like rehabilitation and handling traumas and other serious peace-restorative issues Manipur state need pay special attention. In addition, "peace education" need be provided to those rehabilitated and insurgency-vitiated (innocent) public for expeditious post-conflict restor-ative measures.

5. Intractable Loktak is yet another Achilles' Heel

In a write-up, date-lined Imphal; Feb 20, 2010, under the caption "UG-hunting machines set for May debut", due heavy-handed effort of the Central Government in Manipur Government's current counter insurgency operations is seen.

"Apart from pledging financial assistance on security-related expenditures, the Centre is closely monitoring counterinsurgency measure being taken up by the State Government. As a part of the Centre's thrust to effectively deal with the insurgents and restore peace in the State three hovercrafts acquired by the Government of Manipur from the United States with the Central Government funding the purchase are likely to be brought here in May *(height of the Monsoon)* this year. The amphibian machines are to be utilized in counterinsurgency operations to flush out cadres of various insurgent groups who have established hideouts and operate from the floating biomasses of the Loktak lake in Moirang about 40 kilometers from here. Moreover to discuss present law and order situation with the State Government officials, Union Home Secretary G K Pillai is scheduled to arrive here on February 26. Pillai according to informed sources, had recently held wide-ranging discussion on security issues with his counterpart Brigadier General Phon Swe in Myanmar's Yangon. The source has also confirmed that the three hovercrafts purchased with the Ministry of Home Affairs' financial provision of Rs 5.23 crore would be put into operation from May this year. The machines being acquired by the State Government is said to operate at high speed with capacity to carry 10-20 military personnel with equipments. The said hovercrafts would run on diesel, said the source. It is said that during the course of several top ·level meetings between representatives of the Central and State Governments with the main focus on sanitizing the region's largest fresh water lake from being transformed into a haven for disruptive elements. The officials were also of the same view that most of the anti-establishment activities were being orchestrated from hideouts in the Loktak Lake and as such a unanimous agreement reached to utilize the hovercrafts for effective checking of UG actions by crippling their operational centres. The source further reminded that for maximum impact of the counterinsurgency operations in Loktak Lake and efficient handling of the machines, Loktak Protection Force had been specifically raised by the Government with units of the special force already deployed at four strategic locations in the lake's periphery.

It also underscored that incrcase in the frequency of counterinsurgency operations launched jointly by the police and the Central security forr.es during the rer.ent past is a part of the policy to weaken the UG networks and to gradually strengthen the area domination policy of the government. The development assumes significance as the DGP Y Joykumar had some days back exuded confidence of freeing the State from the scourge of insurgency. Regarding G K Pillai's schedule to arrive here, the source said the top Home ministry official will stay for a night and is expected to engage security officials in lengthy deliberations to chalk out concrete measures for bringing a lasting solution to the problem of insurgency. The visit is being considered as vital and very important specially in the backdrop of the Home Secretary's meeting with the Myanmar Brigadier General, during which the focus is said to be on presence of training camps and hideouts of UGs of Manipur in the neighbouring country. It may be noted that, apart from the presence of UGs on Myanmarese soil, the State Government had been persisting with the Centre to construct fencing along the international border to check entry and exit of armed cadres. Meanwhile, information has been received regarding a policy for deployment of at least 5/6 battalion strong Assam Rifles personnel on the Manipur side of the border out of the total battalions being raised to man the porous Indo-Myanmar border. To effectively check activities of nearly 44 armed groups in Manipur, plans are also afoot to construct 100 helipads along the international boundary. *(The Sangai Express*: 20 Feb 2010: p1).

Of late, the Nagas 'seems' to have acknowledged the saying "better bend than break" as the once much-debated demand for a sovereign Naga nation 'appears' to have narrowed down to the formation of a unified homeland (integration of all Naga-inhabited areas under one single administration) within the Constitution of Indian. Unification of all Naga-inhabited areas was considered legitimate, non-negotiable and formed the backbone of Indo-Naga talks. But India Government has cleverly assigned the task to the Nagas to solve the problem by themselves and refused to tackle the problem together. The response of the neighbouring State was disheartening, if not outrageous . . . Manipur where the Meiteis (Manipuris) dominate and rule the hill people rose up in furor creating a hitch to the . . . talks.

6. Ending a Conflict

The best way to overcome the Taliban was steeped in the recipe of "divide and rule" *(No Time to go Wobbly;* October 15th 2008 *Economist*). With war fatigue clearly visible on both sides, it is diplomacy that is now required to nudge the parties to the table. The negotiation process will be arduous but will be less painful than continuing current hostilities. The mothers, wives, and children of all sides want an end to this war; but are being fuelled by unacceptable "collateral damage" from both side of the border. A Pushtun is a fine and proud person. He loves his rifle and his old wife and strongly believes in revenge, if justly wronged.

7. India's "Core Concern" following Pune terror strike (February 2010)

While agreeing to resume the Foreign-Secretary level talks (25th

Feb. 2010) between India and Pakistan, despite the Pune blast, the official sources admit at "nuanced approach to dialogue (is) being adopted" aimed at building a climate of trust for future talks between the neighbours, but that much will depend on what Islamabad does. India is being "realistic" in going ahead with the meeting, as beyond a point, putting off-for all these 14 months since the 26/11 Mumbai blast — was likely to yield diminishing returns. However, the official Delhi source adds: "However, this [meeting] is not aimed at resolving the outstanding issues. We are fully conscious of the complexities involved in the process and are, therefore, adopting a nuanced approach to the dialogue". (*The Hindu:* 19 Feb. 2010). The sources add that terrorism directed at India remained "our core concern" and the terror strike in Pune put this concern in sharper focus. New Delhi was in fact clear that the talks did not signal resumption of the composite dialogue that was put on hold after the November 26, 2008 Mumbai terror attacks.

> "The Indian focus during the talks would be on terror. If Pakistan has other concerns, we are willing to address them. However, the future course of the dialogue would depend on how they respond to our concerns", the sources added. There was a legitimate reason including a sharp domestic reaction, for suspending the talks in the aftermath of the Mumbai attacks. While this strategy has paid good dividends, today the situation had reached a stage where "not talking" could become counterproductive, more so given Pakistan's own vulnerability in dealing with terrorism.

The Af-Pak situation became quite different from what it was a year ago. "The core establishment" led by the military, had strengthened itself *vis-a-vis* the civilian population and wanted tension on the Indo-Pak border. "The establishment wants to rally the whole of Pakistan behind it and nothing works like a bombing . . . So if we were to have an instinctive reaction, this would be playing into their hands", the sources added.

Referring to the provocative statements from across the border, among them the February 5 warning of Abdur Rehman Makki that

Pune could be a future target, the sources stressed the importance of dialogue in cooling the temper and restoring normality. "We need to be sensible in the way we approach each other as we are both victims of terror. If anything, we need CBMs (confidence building measures) on terror. Whether we talk or not, terrorists have continued making attempts. The way you deal with this is hardening yourself internally".

8. Redefining 'Conflict'

Conflict indeed covers anything opposed to accord, agreement, harmony or symmetry of thought and action. It arises from an incompatible wish or need; or hostility in thought or 'dashing principle' with a view to overcoming, by use of force, the 'other' conflicting party (actual war); or preventing, interfering or intervening, through use of force, between two conflicting parties, in short a 'war to prevent war'. It therefore covers both the pre-war (leading to a possible fight to an ongoing struggle) and the post-war situation (encompassing the resultant distress like war reparations, fresh military build-up etc). Etymologically speaking, it need also cover in its penumbra all its impact on social attitudes from traditionalism to modernity as embodying changes caused by conflict (preparation by Hitler-led Germany prior to actual outbreak) in human society. Not to be outdone by war-mongers, the peace constituency effort to 'prevent war' ought to be encompassed. On that score, post-modernity may indeed make do with a 'conflict', redefined as violence-tree or non-coercive effort and means to bring about changes in society.

Any such change one sees in. history or finds in human exterior is thus directly or indirectly attributable to conflict or dissent. But, like anything else, even this 'conflict' — defined as the very genesis of 'change' — has also been subject to 'change' or at least adjust and evolve (morph) against the powerful current and forces of time. And its latest format, since recorded in human history viz. all-out and most offensive 'modem warfare'-the most horrendous version of 'conflict' - stands more or less effectively abolished all through the latter half of the 20th century, But the early-21st-century has found a still fiercer format: a mish-mash of stealth and innovation to cause maximum 'spot on' devastation and trauma. This *jihadi*-attack may be deemed as a perverted version or perhaps an appendage of 'insurgency', in the sense that it has much greater outrageous impact on humanity at large. The worse part of the story is that unless one and all do willfully and conscientiously abhor 'conflict' by practicing non-violent (peace-loving and presumptively 'peaceful conduct'), the world will never attain 'positive peace' by abstaining conflicts in thought or action. And no amount of deterrence can hold back a desperate (fundamentalist) person from resort to 'terror', unless he himself sees his own thought and action wrong, and even if he does not, there are none else, sponsoring the jihadi spirit or action and funding it.

Some relevant questions arise: whether 'conflict' has seen greater desperation on the back of the so-called "mindlessness" —the inverted form of 'reason'. But, here too again one finds oneself in a tight spot to make a 'clarion call' in the 'murky' character of the present era, as in a murky weather (poor light) it is 'never easy' for the umpires to decide whether to start a crucial game or not. More so, when the current humanity has seen affluent and also emerging nations as well as the so-called unprivileged or the desperately poor 'Third World' struggling to make do against hunger, famine, poverty and poor health. Certainly, the desperation makes one recoil and fight with one's back to the wall. That is the time when reason is banished from the desperate fighter's mind. Even an acknowledged sportsman may find a thin demarcation line between gamesmanship (win for the sake of winning) or tor the furtherance of the spirit of sport. After all, why sport in conflict. There is a point in banishing war–at least for seven decades since World War II–and possible annihilation of the vanquished by the victor eliminated. So, there will be no defeated nation to seek revenge through another war. Yes, 'war' has been redefined. Its genesis 'conflict' has been re-paradigmed. The possible cause of any war is now stumped out before the actual outbreak of the serial war. Likewise, there will be no wasteful use of scarce resources for belligerency–hence divertible as 'spin-off' to peaceful purposes to help the needy, the desperate elsewhere. What proved effective was the reining in of the use of 'force' the inverted form of 'love' (for another colleague).

Over and above the annihilation, traumatic stress, the killer himself suffers a setback once the killing instinct is accomplished. The very act of killing which is as fundamental to a fight over game, or to war, or to most violent terror-strikes. Yet it is also the one thing many combatants (including the suicide bomber) avoid citing both before and after killing, out of fear, shame, guilt, or deep fear that they will be misunderstood. The *jihad* of course lives in an altogether different world. He is a mechanistic perpetrator of violence (terror) with the aspect of his becoming an instrument of killing others (the 'victim' prognosis reduced to near-zero). Thus for a jihad it is always 'win-win' and never a chance-game of 'lose-gain' true for a combatant–who may get killed and become a 'victim'–that residual transient feeling where he survives the encounter, and carries the post-traumatic stress, which almost always leads war veterans to post-traumatic stress disorder, alcohol abuse, anger and relationship problems. How the stigma and shame of a war veteran (combatant) can be reduced to save his future life has now become a new field of medicine in the wake of the war machine's emphasis on causing maximum annihilation.

Transcending ages is another ingredient of conflict being 'intrinsic to mankind' and his story. Conflict is circumstantial, consequential and incumbent upon his own existentiality, even if not a direct creation of his 'self'. If in the first instance 'I' as the self protrudes by indulging in an act of hunting (farming), I might consciously or otherwise overstep on the 'other'. It is thus the necessary evil, seen and experienced right from his childhood among mates in schools, homes till his ultimate self-recluse, or ultimate retirement. Only two formats of 'Conflict' are fundamental: incumbent and non-incumbent. While the former is one that follows my own action (protrusion), the other arises because of others' actions (protrusions). Naturally, the more competitive is one posited with, the greater will be its recurrence. The turn of the 21st century saw greater resort to terror-strikes. Since then the prime terrorist target viz the US has re-strategized its Homeland Security in a rare mammoth effort and its recurrence is brought down almost to mere threats. Obviously, terror-strikes could also be minimized - more like "war" has been avoided — if the fundamentalist could be 're-integrated' or politically empowered and rehabilitated. If 'Taliban', the so-called US-creation in its superpower race against USSR in Afghanistan. And in the second decade of the 21st century India found itself as the next terror target and a similar strategization was under way for having stringent vigil through the Internal Security measures to reduce terror-strikes. How inextricably involved is humanity in the genesis (almost a ball-game) of 'conflict', and even 'social change'.

9. Liberating the Liberated Zones

In the year 2004 South Manipur was known to be a strong hold of the Valley Based Insurgent Groups (VBIGs) and considered to be a "Liberated Zone". The line of Liberated Zone had virtually been extended to entire Chandel district (less NH 39 Pallel to Moreh), entire Churachandpur district up to Jiribam less NH 53 (Jiribam to Nungba) bought under its ambits subsequently. Fights between Naga UG groups and VBIGs had also touched Tousem sub-division area of Tamenglong district when central government realized the complicity of the situation. IED blasts/ Plantation of IEDs (Improvised Explosive devices) became common and questions had to be raised by State members of Parliament in the Lok Sabha. The hue and cry was taken note of when Th. Sangliana, IPS turned MP from Bangalore who otherwise hails from Sub urban areas of Mizoram touching Hmar areas of Manipur made it loudly audible with thunderous pleas in the Parliament, why Hmar people were alienated by the Govt. of India giving free hand to the VBIGs i.e. to shut innocent people in their homes as they were scared of going in the fields and jungles due to fear of planted IEDs. About 50 to 60 had already lost either limbs or life 800 to 1000 Hmars had fled to Sawkerdai adjoining sub division of Mizoram for shelter and safety. This situation had arisen because for 20 years or more than that the deployment of police was most negligible in these two zones. If it was, it was a handful of civil unarmed police personnel only and the entire area was officially known as POLICE SECTOR with literally no police presences.

Civil administration had gone defunct with the entire police outpost virtually closed down. Chakpikarong sub division of Chandel district was a predominantly Kuki dominated area having rugged terrain, rivers interspersed with deep gorges, riverines and mountains. Loktak Lake is like a miniature sea. It is the largest fresh water lake in the North-East. The entire lake was covered with floating masses called "Phumdis". Many fishermen and their families live in small huts on the floating phumdies. Traversing through phumdies is extremely difficult unless cutpaths are created by manual clearance. At places boats had also to be lifted and redone on available clear water surface for sailing through. Thus these two zones had made it a perfect setting for insurgency. There was no civil administrative machinery in place as none of the government officials wished to serve in these two zones. The villagers supported the groups out of fear, providing them, administrative support and an intelligence network.

The only security forces in the district were AR Battalions that too on National Highway 39 to keep communications open. They were unable to make headway towards areas around Sajik Tampak due to the massive UG presence, and moreover area falling under police sector. UNLF, MPLF, PREP AK, KYKL and PLA were the major groups operating in the area. The UGs had a free run in the sub division with their permanent camps perched in the jungles, temporary satellite camps in the mountains/re-entrants with administrative and intelligence backup from villages and across the IB respectively. In Chakpikarong in early 2004, it was the writ of the UNLF and MPLF which ran. Similar was the case with Loktak lake and other interior areas of Churachandpur district up to Tipaimukh Sub- division. Area Chakpikarong became ideal place for training and veiled operations of the UGs and Loktak as rendezvous for summoning senior Government officials and others for various types of negotiations on extortions, cuts from Government funds and coercing some to obey their dictates on employment of their sponsored candidates and so on. Chakpikarong also afforded safe routes to go into Myanmar and back area beyond Singtam short of Chakpikarong became the cut off point for state forces to venture beyond. 2004 Parliamentary elections approached. There was surmounting threats from the UG outfits on some candidates not to contest. Even one important leader Th. Chaoba Singh former MP–a BJP candidate had to withdraw from the contest. The Election

Commission of India wanted the conduct of elections as free and fair, so Manipur Government Ministry of Home Affairs wanted Army to be deployed prior to elections to neutralise the interference capability of the UG outpts. Hence the Army was called in to check their free run in the Chakpikarong sub division. The 44 Mountain Brigade was moved from Chaprakata in Assam to Chakpikarong sub division in Apr. 2004. The Brigade at that time consisted of three Army Battalions (27 MADRAS, 9 RAJPUT and 7 SIKHLI). The Brigade was deployed with the HQs at Chakpikarong and battalion each at Sugnu, Joupi and Sajik Tampak. And 'All Clear' operation was launched at Chakpikarong and Loktak simultaneously with 44 Mtn. Bde in Sajik Tampak and 73 Mtn Bde in Loktak area to achieve the national aim of conducting trouble free general elections and blunt the striking powers of the UGs. Wasting no time, the units commenced their operations with full vigour. Recce of the valleys of Sajik Tampak and Khaokhal was done to identify and target the terrorist's strongholds. The initial operations faced stiff resistance from the terrorist groups, with many patrols getting involved in fire fights. With the increase in casualties to the terrorists, the pattern of stiff fights gave way to retreats as terrorists avoided contact with the Army and withdrew to their deep jungle hideouts.

The task ahead was a complicated one to take on. The UG had mixed up with local population. At strategic entry points to the villages 'Sangars' were made to be used as firing bunkers. There was a camp of NSCN (IM) called ***'also taken note of'*** as per Ceasefire Monitoring Group analogy at Phungchung which was supposed to be effective within Nagaland only. And there were no specific orders with the GOC 57 Mtn Div at Leimakhong and IGAR(S) at Mantripukhri, and also with Govt. of Manipur how to deal with them. This NSCN (IM) Camp was not interfering with any security forces activities actively or passively but its cadres were always creating some or the other trouble with Hill and Valley UG groups for domination/coercion of locals do and their ideology. It took the turn of force a changing ethnic loyalties of small and weaker sections from Kuki's to Nagas and Nagas to Kuki etc., with VBIGs pushing their own agenda through and through. Such activities went so intricate that authority of chiefs also came under erosion and pleasing or displeasing of community leaders to pull them to their ideology carried on and on and many lost their lives in this game. Moderate group of people were always under their scanner and threat if they refused to cooperate with them. Above this NSCN(K) was also involved and used by groups other than NSCN(IM) to grind their own axe. Prior to induction of Army or Assam Rifles, BSF had given a failed trial to penetrate into Chakpikarong stronghold of VBIGs and they were forced to withdraw from Singtam itself with loss of one officer and one constable dead. With this backdrops, 44 Mtn Bde were to first to disentangle civil population from the armed UG groups, then move in to jungle areas and finally to fight if necessary to force their surrender but definitely to throw them out from their strongholds and cripple their armed strength so that elections could be conducted trouble free. As task not so simple, of course! Above that to ensure no co-lateral damage occurs.

Army employed many innovative types of ruses. After sunset at unsuspected and at staggered timings, troops used to resort to continuous illuminating fire with maximum quantity of the training ammunition live and blank both of Infantry light and medium weapons including of high trajectory. For this areas used to be reconnoitered during day and selected only which were population free including animals but in close vicinity to village(s). Check gates like at Shaluk, Chakpi, Glangei were established to control movements of locals. The rice was allowed to be carried only for 3 days consumption to be renewed on regular basis. Water and medicines were supplied to villagaes in Citu. By Aug-Sep 2004 Bde was successful in creating an efficient intelligence network and gain confidence of the villagers. Opeartions were extended into the deep jungles. The Brigade shattered many pre-conceived notions of Army operations, by following unique and unconventional methods. Some of the shockers for the terrorists were the active operations of the Army even during monsoon periods, cross country movement, jamming of terrorist's radio communications and ops at unsuspecting hours. This created caution in the UGs and many (including the leaders) moved out to Myanmar. Camps were busted with huge recoveries of weapons and ammunition. A special mention needs to be made about the operations conducted by 27 MADRAS at Thergotang Halkeip which resulted in the recovery of 59 AK-47s, one LMG and one mortar and also by RAJPUTs at Sugnu in which 61 AK-47s, one LMG and one mortar were recovered. Relentless operations were carried out in the villages of Toupokpi, Shaluk, Mombi, Joupi, Hengbhi, Aigi, Aishi and Bather. The major camps of UNLF at Sehlon and Phiphenkot were destroyed. By Sep 04, a focussed drive was undertaken by the Army to completely sanitise the area till the IB for which teams of 21 Para SF were also inducted to boost Bde's fighting capacity This led to surrender of many terrorists who were unable to bear the hardships due to constant fleeing from the Army and unavailability of administrative backup. The SDO's office became functional again with the return of SDO to Chakpikarong, and decks were cleared for the civil administrative machinery functioning to put their acts in order. Army actively engaged itself in the rebuilding of the area with construction of new roads, providing medical cover to the needy, carrying out pre-recruitment training, construction of football grounds and also by arranging free meals for internally displaced persons. Slowly and steadily, normalcy returned to the sub division. The people of South Manipur had their faith restored in the nation. The efforts undertaken by all personnel of 44 Mountain Brigade and 73 Mtn Bde were praised by one and all. On completion of its task, 44 Mountain Brigade handed over the operational responsibility of Chakpikarong to the Assam Rifles Battalions in a phased manner and moved out to Pailapur in Sep 2005, to face a new challenge in Tipaimukh Sub Division. With move out of complete Brigade and plus, VBIG's cadres who had crossed over to Burma judging thin deployment of Assam Rifles sneaked back in strong batches even with lethod bombs and other sophisticated arms. Since the entire area was entrusted to battalions, of AR, the AR troops had to face major blunt of the UGs and some causalities' in deaths and injuries were also suffered by them. Therefore a complete DIG HQ with full quota of new battalions and SSF troops were moved in to Chandel district. Sajik Tampak (Complete Chandel) is now looked after by 26 Sector Assam Rifles who are suueccessfully engaged in Nation Building activity with hand in hand of civil administration. On the other sector 73 Mtn. Bde achieved successes in Loktak Operation. Motor boats were used to penetrate in to the UG areas of hiding and at places heavy firing ensued. In one of the major clash in water UGs lost 10 to 12 cadres though media confirmed 8. Army also lost a couple of its other ranks. But UGs citadels inside Phumdies and complete control over population was shaken. Loktak lake Brigade since then is continuing its operations against UG and steadily blunting their strikes and hide power. UAV (air craft without pilots) say unarmed aerial vehicle, excessive use of night vision equipment were put into use. Now recently three Hovercrafts for Loktak lake Operation have been inducted and Government has cleared Loktak Lake of dense phumdis at an estimated expenditure of 300 crores of rupees. Areas both Chandel and Loktak are rapidly getting new look under IGAR(S) and GOC 57 Mtn Div respectively.

10. Viable ways to safe secure Northeastern Regional Security

There is no denying the fact that Myanmar has viable potentials to help India achieve safe corridors in the SE Asian Zone because of the following facts:

(a) North East has shared histories of British rule, co-moving, economic dependencies and same ethnicities sharing common International Borders earlier through one province i.e., Assam with Burma and now with Arunachal Pradesh, Nagaland, Manipur and Mizoram with length of IB neither changed nor stretched/ altered

(b) Manipur, Mizoram closest to the sea offer viable route accessibility to sea in an important aspect of global requirement.

(c) Sittwe sea port in Myanmar a nodal point for oil availability falls enroute to proposed trilateral trans- Asian Highways, In the Pipe line project, lies its crucial importance. It becomes point of common interest on oil and gas requirement for both India and China. China has achieved connectivity to Sittwe from Yunnan Province through land route which will cut short its compulsorised Navy voyage of 3000 km plus in addition to avoiding the use of Strait of Malacca - a tortuous and unsafe route and seeking access to the Indian Ocean through Myanmar via Bay of Bengal in the longer run. India also needs Sittwe connectivity by land route and sea both. Kaladhan Multi-nodal Transport Transit Project on completion will facilitate this and India is also involved in its development. But it needs more pro-active and fast moving diplomatic cards to invest in Trust Bank with Myanmar Government as Chinese did. They pulled fall out of western nation's sanctions and Indian idealistic stance of foreign policy decade back to their advantage and extended quick credulous hand needed at time of their dire need by moving financial and military help to improve their land routes and some sea and airports too. Ports up to Karachi like Gawador and Sri Lanka like Hambantota were also developed. Imagine its impact during war if it breaks out and Chinese push in trading activities with Bangladesh, Pakistan and Myanmar off the war times. Since India is on trade pact with Myanmar, its thrust into India's NE economy cannot be under looked.

(d) India must invest more in trust measures to win over confidence of Myanmar rooted in Buddhist Thought common to three nations - India, Myanmar and China and moderate Myanmar to liberalize their democracy to suite their requirement of course but more people friendly. Two giant nations of Asia enroute in close race to outshine in world economies are in race. By mutually basing age old faith in religion, culture and philosophy both India and China can grow faster and get closer for which Myanmar can play a greater role. It will improve its economies too by acting as conduit for India and China and other SE Asian nations. And it can be a stepping stone for ASEAN Nations to establish common market like EU (European Union).

(e) Most importantly all the three nations face common threat from Islamic fundamentalists in one way or the other under Al-Qaeda horoscope. If the trio- India, China and Myanmar use their cultural ties and trade potentials to positivities, the threat of Islamic fundamentalists and of NE/Myanmar's insurgents can be neutralized.

It is an historical fact, Hill people of NE and Myanmar lived in penny pockets, away from main civilizations generally represented by valley people due physio-geographic compulsions. Tribes to tribes affiliations governed by indigenous thoughts and beliefs with interfaith on localized trade and commerce on need and feed basis, their fates sealed under feudalistic grip of the Chiefs who were further governed and influenced by the diverse systems of governance ranging from Rajas, Avas to British Administrators and always suffered whipping by strong rulers Due to grand partition of India and Pakistan (one now Bangladesh) and carving of a separate country Burma in 1947 and 1948 respectively, common people living along divided lines of international boundaries expected a system of hereditary nature of independence and localized political governance of their systems, but it did not happen. It is also a fact under consolidated plan of village states turned into organized state it was just not possible to work. A new type of isolation, fear psychosis enacted out of the militant activities came to their kitty especially for those who live either side of Indian Borders (IB). Modernity affected by imposed religions of Christianity and Vaishnavism affected their life style; whereas all modern religions are based on primordial philosophy which is righteous and ethical deeds. Independence which should have brought happiness and satisfaction brought hardened systems of living for them. Boundary pillars considered to be benchmarks of knowing physical extends of nations came to be taken as dividing the common and akin ethnicities either side of the BPs (Boundary Pillars) to their dislikes, disdain and pain. Physical extends could not transcend the geographical extends. Move of locals either side of the IB within age old traditions and permissible limits i.e., 20/16 km either side w/o permits during the British times had to be discontinued due move of militants to Myanmar soil and security vigilance tightened by the SFs manning the IBs. No efforts have been put in by both sides of government to put up police posts either side at affordable distances/proximities other than police posts at trade centers i.e., Moreh/Tamu in Manipur and Champai in Mizoram.

It has caused not only distancing them but also separating them from one another but also separating them from their own kith and kins and many people losing their lives on interception at IBs by the SFs either side and also inter intra insurgents moves. Myanmar since practices militarized democracy so their Regiments, even resort to burning the villages who change their religion from Buddhism to Christianity as and when their ops against militants of any hue or otherwise necessitates Naga Insurgent Groups on operation within Myanmar have one strategic group to popularize Christian faith with locals hence conversions. Areas in Heimi control and under Kachins Independent Brigades have thus become dens for drug trade and even pronounced training centers for almost all the NE insurgent groups with exception to none. 1643 km of IB with Myanmar has become an area of ***'no man's land'*** for negativities. Entire areas are porous, difficult to be fenced and if someone can walk along the IB, will find it dark energy sapping which many a time have triggered disaster with locals at hands of Burmese Regiments and utter unproductive where as if both nations put up common projects with common interest and mutual trust, the areas can be turned into highly productive and growth oriented not only in eco and cultural tourism but may be helpful in shoring minerals, timber, wild life and gas exploration too.

India and China are part of the 14-member, 'Friends of Myanmar' group and Thailand was instrumental in creation of this group. Thailand services can be beseeched in initiating common projects with Burma. The Kaladan Multinodal Transport Transit Project provides a great opportunity and will give India access right up to Sittwe port. As Myanmar progresses economically, political changes would develop unseen to sustain its growth and it may be possible for Burmese to switch over to liberal democratic governance in a win-win model for all stakeholders which will affect Thailand also at that stage. Presently there is an urgent need of opening Manipur window in its ***'Look East Policy'*** for art, culture, dance and musical exchanges with China through Myanmar. In the second stage experienced, matured persons from China side to be invited to visit Arunachal Pradesh sub divisions one by one for them to see and analyze the cultural proximity of tribes/communities Arunachalese and Tibetans if any other than Taiwang. There is a contrast in living styles and customs in those tribes dictated by geographical divides. It is worthwhile extending a 100 feet wide tarmac road along 1643 km IB with Myanmar to extend tourism facilities, gasoline explorations, with health centres, motels, workshops and civil police outposts to facilitate people's move either side more freely though under legal and security check.

Delhi needs considering the following

- Internalize all blames on our systems. Neutralize bigot fundamentalists systematically and avoid passing buck for everything sloth to Pakistan. Muslims and Hindus must contribute for national peace, harmony and expand interfaith space and for this judiciary be made more strong.

- Explore potentials of Chinese-Indian co-operation to mutual uplifting of economies by commonalizing communications-trade, water gas exploration and art- cultural exchanges and get out of the Chinese 'haoo-ah' created by various analysts and so called experts and retired seniors who have never seen the IB from close angles. If both sides agree not to interfere 10 kms either side except at most unavoidable points, with mutual consent, the border problem will never be a bug and a matter of dissent as there is no other solution to adjust borders. This problem is realized by only those who are inhabitants there, who have no idea, what is the security risk around them. That is their natural living zone.

- Bangladesh creation was unavoidable as a political necessity though strategists at that time (1947 and 1971) never realized it would be economic partition too. People of Bengal– Bangladesh have common - heritage multiple that is religious philosophy, art and cultural. Even freedom struggle witnessed common sharing, if so then why today, any mutually effective developmental projects are viewed with suspicion? Take example of Tipaimukh Dam and establishing of new sea ports, why politicize it?

- Let our all connectivity impacting NE and Manipur be in tapping water resources etc, free flow of trade to get through as normal and at a faster speed.

- Crackdown on corruption now in motion must be accelerated.

And at Manipur level in particular:

- Rehabilitate internally displaced tribesmen at their original places. 15000 Anals are crammed up in and around Chakpikrong (Chandel District) since 1992-94 due then Naga Kuki conflict.
- Kuki's and Nagas be persuaded to restore each other's lands wherever grabbed like Joupi (Tamei), Tamenglong and Makhan (Motbung) area and other places well known to the government They are also 1992-1994 conflict results.
- With peace all round, UNC and KNO with respective elected leaders and church, leader 'must sit together and declare forgive and forget the past bitterness once for all and then Kukis be asked to drop annual black day observance for healthy growth of society.' The veteran politician and living legendary figure Shri Rishang Keishing H'ble Member of Rajya Sabha had done it for bringing peace between Paites and Kukis at Imphal by a feast hosted by him which worked well.
- Put a stop to make Manipur as a destination point for dumping stolen cars from other states. Drugs and Narcotic Act needs a relook. If product of the prohibitory items like poppy, marijuana, heroin is fetching attractive returns to growers/dealers and somewhere it is used for medicinal purposes/otherwise, like South Americans getting cocaine out if it, the Central authorities may think of regularizing it on a restricted scale with permits so that State can earn revenue out of it. And slowly but steadily it can be eradicated with introduction of alternative crops in fruits and orchids, special grade bamboos and herbal plants as Manipur soil is extremely rich and favorable backed with salubrious climate. Dubious businesses of all kinds like quick doubling and multiplying money (which largely rob the investors) must be banned in total. Such activities advantageous to only a few erode national disciplines which acts as route to other nefarious activities effecting internal security. Sound internal security systems are

of immense importance contributing to external security and integrity of the nation.

- Make civil administration deptt functional up to EAC level, (Extra Assistant Commissioners equivalent to Tashildar in main land) in situ. Old fashioned posting order at civil adm outpost Chingai with HQ at Ukhrul i.e. Distt. Hq. be done away with.

- Place one IRB Bn each with appropriate civil police and ED (Enforcement Department) up to SDO/EAC level so that all other deptts can function there with safety and security. Local Committees from responsible citizens be formed to take care of their safe stay. Village chiefs and council members can play a good role.

- PHC/Subsidiary health centers must have doctors/ para medical staff with adequate medicines stocked and ambulances as a national priority.

In fine, the synergy level obtained between civil police/SFs and civil administration be enhanced further. Current achievements are possible only because of safe security environments so created by this trio.

GLOSSARY

active learner: a peace-activist in 'conflict-resolution module' ultimately aiming at a just, peaceful and harmonious society generating creative development (rather than destruction).

active learning: a serious learning to have an in-depth knowledge and their potential benefits to mankind.

active learning methods: which` allow teachers greater flexibility in adapting often-problematic curricula. These methods also challenge traditional authoritarian classroom relationship and a banking-style of learning in which the teacher is expected to be the 'expert' and students 'receive' information from them.

active teaching method: an umbrella term covering a wide range of teaching and learning methodologies designed to engage students actively and critically in learning. In contrast to a traditional one-way, banking-style of education, active-learning methods encourage students to engage in two-or-multiple-way dialogues with the teacher and each other. Active learning methods often use drama or physical movements to make learning more memorable, interesting and enjoyable.

adversarial relationship: the near-explosive character shown by two conflicting parties in the penultimate or final stage of conflict; the penultimate stage to which a party (in a conflict) can escalate his discord animosity with another.

agenda-setting (talk): initial meet between representatives of two conflicting parties exploring the scope, nature, purpose and coverage etc of talks.

alternative conflict mapping: mapping of a conflict situation from the stance of peace building, in terms of non-conflictual terms by using non-conflictual complexes of identification which can create congenial conditions for all-round amity and regenerate peace for facilitating meaningful non-conflictual interaction.

Anti-chauvinistic education *(or examining hegemonies)*: aims at a broader, systemic look at discrimination, rather than anti-chauvinism education which focuses on chauvinistic attitudes and behaviour. It would also examine dominant ways of thinking that are chauvinistic or discriminatory in their implications.

audacity of peace hopes: on the optimism of hope the Bible says: "there ariseth a little cloud out of the sea, like a man's hand"; the little hand is destined to rise higher and still higher with astonishing celebrity and will explode with a deluge of rain on the parched earth; even the Taliban tangle - hitherto eluding peace ever since the colonial days when Britain failed to impress the Afghan tribals down to Russian and American (NATO) still stuck up on 'reintegrating" and 'reconciling" the Taliban in the Afghan national mainstream.

bad peace: a form of negative peace, wherein peace is maintained (make done), neither lasting nor sustainable.

bedrock of sanity and amity: the foundational and the ultimate irreducible adjunct of human milk, from which any emotional humanoid could have raised, and raised his/her level of incompetence (trust deficit in the existing institutions)

banking style of learning: traditional authoritarian classroom relationship where the teacher is expected to be the 'expert' and students those who 'receive' information from teachers.

bilateral cold-war time-wrap: an indefinite .

biological reductionism: the currently empowered group (political leaders) are not taking insurgency as the so-called 'at-stake' template of the 'rational' or strategic human needs dimension of a repressed and dissatisfied group. Counterinsurgency is thus viewed as attempt by those in power to 'biologically reduce the group. Hence, counterinsurgency is perceived by the insurgents (aggrieved party)' as 'biological reductionism'.

capacity-deficit: state of counterinsurgency apparently revealing explicit inability to cope with the ongoing momentum of insurgency

citizenship: *such* citizen as possessing an 'enhanced character, more sensitive to the needs of the community, more competent and committed in their ability to contribute to society, and more civil in their thought, speech and action;' in addition citizenship involves a second set of ideas and values, which occupy centerstage in the *liberal* tradition, linked to the ideals of parliamentary democracy in which citizens elect representatives to look after their interest, and thus converging to the *'rule of law';* T H Marshall distinguishes three stages: (i) civil (or legal): right to own property etc; (ii) political: right to suffrage, etc; (iii) social: access to social goods welfare services; For a conflict-ridden society UK's official Crick Report merely puts down social and mom responsibility, besides political literacy: broken down into (a) care for others; (b) premeditation and calculation about what effect actions are likely to have on others; and (c) understanding and care for consequences which speaks for themselves; (for more on peace and CR skills, see Appendix III: Crick Report).

citizenship education: discussion of what it means to be a citizen of a country; who belongs as a citizen, and how they are represented in public institutions; a complex term whose definition seems to change over time from Greek 'citizen democracy' down to an "obligation to identify 'self' with the 'other'; 'self' as 'other'–and finally to identify with one's community and "hold its interests as dear as one's own (should they conflict), no matter if that community is a town, city, state or country." (L Massey: *On the Origin of Citizenship in Education: Isocrates, Rhetoric, and Kairos*, 1997 <http://cicero.smsu.edu/journal/articles97/massey.htm>.

civic order/society: emotions and formation of social identities; dictatorships, democracy; but with paramilitary forces, community policing to warrant human rights, and with statutory bodies to conduct elections to elect people's representatives.

civil society: the bedrock (mainspring) of society which holds out, either unpolluted by any passing (violence) anti-regime stance, almost always seeking neutral (non-violent) forum and extending its cooperative hand to similar voluntary societies across regions, nations or the globe (and not with non-civil societies).

civil society education: an umbrella term for education aimed at teaching about democratic processes and strengthening elements of civil society including effective voluntary and nongovernmental sectors and a free press.

cognitive social psychology: approach for an individual finding psychological security and self-esteem by identifying *Self* with positively valued group, *in comparison* to perceiving self as *'not like'* certain negatively-valued group; thus during intense conflict as much psycho-logical distance as possible has to be maintained between in-and out-groups (Turner J C: *Rediscovering the Social Group: A Self-Categorisation Theory:* Oxford 1987).

confidence-building measures (CBM): mechanism intended to restore trust or confidence among public on either the volatile nature of a situation, or breakdown of law and order, as during a conflict.

commitment lag: apparent shortfall in commitment of, say, law-enforcement agency during violence, conflict etc.

communication breakdown: a stage of hostility between two parties, when the other party is rigidly stereotyped in negative, hostile ways, and no possible solution to the conflict seems mutually acceptable (Fisher R J: *The Social Psychology of Intergroup and International Conflict Resolution:* New York: 1990).

community building: voluntary effort at the community level to overcome prejudice and *insure diversity* by emphasizing on commonality of purpose and areas of interest to sustain lasting and positive peace through harmonious living (prevention of conflict); but it takes an uphill task once conflict breaks out.

community policing: voluntary effort to prevent violence (crime) at the community level, with or without the help of the Police machinery.

composite dialogue: peace talk over composite (wide range of) contentious issues which have sparked (or are about to spark) off a conflict.

conflict: a normal and natural art of lifeways; but divisible into (a) violent conflict; and (b) nonviolent conflict (competition) among participants as in sports, academic event, ,which will not necessarily lead to violence.

conflict: *Complex Political Emergency (CPE):* a multidimensional conflict crisis with large scale human rights abuses and profound human suffering with the conflict rooted to historico-political reasons; and the state collapsed/contested; part of whose postconflict recovery involves reconstituting the state (Cliffe, Luckham, 1999; *as quoted by Nick Lewer* in *University of Bradford Working Paper* 3; p. 4).

conflict *(International Social)*: a concept defining interstate (international) conflict or something sprawling between the two; often deep-rooted, protracted and considered intractable; *(Ibid Nick Lewer: p.4)*

conflict *(intractable)* : deep-rooted stage of the conflict, where the behaviour of each party to achieve its goals is perceived/experienced as threatening/damaging by the other (Kriesberg L; *et al: Intractable Conflicts and Their Transformation:* Syracuse 1989).

conflict *(protracted social)*: prolonged, often violent struggle by communal groups for such fundamental needs as security, recognition, and acceptance, fair access to political institutions, and economic participation (Azar, E: *The Management of Protracted Social Conflicts:* 1990).

conflict dimensions: psychological, behavioural and structural *(Mitchel 1981)*

conflict history: timeline of conflicting events: World War I, World War II, cold war, nuclear arm race; residual side-effects like terrorist attacks

conflict map *(alternative)*: another possible explanation of the identity crisis, strongly suggesting that identities are only meaningful in terms of interaction that is their source; some individual may maintain conflictual attitude towards some members of a group, and non-conflictual attitude towards other members of the same group; hence did not feel a crisis of identity; and this was because those identifications were negotiated in terms of these specific, interactive experiences with particular 'others'; but once the quality of those interactions changed, or where experience of non-conflictual interaction was denied to the individual, the identification of *Self* and *Other* was re-negotiated; using such a roadmap it becomes possible to systematically identity key historical and contextual influences on development of the conflict; to identify the parties; list the issues; explore the conflict dynamics (emergence, polarization, escalation, psychological factors such as stereotyping and prejudice).

conflict-mapping: to ascertain the multi-factoral and dynamic process of conflict, academia started research on structure and context of complex conflicts from both subjective and objective perspectives; if the particular social-psychological perspective on conflict "is as informative and constructive as is assumed, or whether alternative perspectives might have as much or more to other"; but such mapping is highly subjective (transitory) and its transcendence over time after the conflict situation is negotiated; its transient nature is also because it is not 'objectified by the knower'; in other words, far from existing in a fixed form, waiting to be observed or known, the (social) reality is constructed and negotiated intersubjectively; *(for more detail Appendix I: Conflict Mapping).*

conflict-mapping *(psycho-sociological)* : a conceptual map to be used as guide to conflict resolution; a psycho-sociological mapping of the conflict in Bosnia in general and in Mostar in particular carried out in terms of the crisis of (psycho-social) identity crisis ensues when any perception of self in the 'enemy' other (or of the other in the self) tends to threaten that gap and, much more like a child, one would have a regressive yearning; as a result, the 'enemy' is *'dehumanized'* in order to account, both, for the enemy's hostility; and also to justify the worst excesses of the in-group's behaviour towards that enemy; however both the in-group and the out-group are somehow alike in the way they behave towards each other; inescapably pre-eminent is the fact that the *Self* and the *inhuman other* are similar. Further this continual threat to the Self drives a dynamic of enmity.

conflict resolution: such an epic event cannot happen merely because two conflicting parties (leaders) are meeting and talking, but more significantly through a powerful (sustainable peace-oriented) awakening among the masses on both sides of the divide; the best example of conflict resolution is where: *'erstwhile adversaries'* could become *'stake-holders'* in *'a shared enterprise'* by rendering' *obsolete their historical antipathies and autarchic mentalities.*' France and Germany, Germany and Russia, and Turkey and Greece were locked in deathly embraces one way or another in modem history (the inter-War years) more or less now the two umbilical nations India and Pakistan. By creating 'creative ideas and initiatives,' Talibans need be empowered, when and when only the 'Afghan-ness' will descend in the Taliban. The only problem remains: who will bell the cat? Else, 'who will set the agenda?' and whether all issues (including any minor ones which may snowball soon, like water-sharing of the Chenab between India and Pakistan) have been taken care of ?

conflict resolution *(positive)*: an integral part of peace education; that *conflict* is a normal and natural part of life since childhood, whether in family and schools or otherwise; if school education onwards the children are taught or become convinced that conflict (name-calling, gossip) is neither (necessarily) negative nor will it lead to violence, then (in manhood) they will deal constructively with conflict and hopefully in a way that will transform participants and teach them something not only about the *other's* perspective but also about their own;*(also see citizenship education)*.

conflict resolution *(in Schools Programme: CRISP)*: inclusion of (personal) conflict resolution (now a normal and natural part of life) in curriculum for young students has been adopted in UK; in fact, there is a fair possibility that the child undertows violence into manhood as a result of (schools) conflict. *Whether actual or imagined*, violence is since childhood viewed and imbibed as the appropriate method of dealing with discord (conflict). The CRISP survey finds that 81% of students resort to name-calling — often a catalyst to violence — every week. Half of the teachers believe verbal violence results in real violence and fights, even students agree with teachers that they need conflict resolution skills to stop name calling, gossip and actual violence; The British survey report may be skewed, viewed from its multi-ethnic population in late 20th century; however it will reflect a much more realistic replica for conflicting societies in Middle East, South Asia or Manipur for apparent reasons; CRISP applies to both imagined and real conflict situations.

conflict resolution *(skills)*: analysis of nature and origins of violence and its effects on both the victim and the perpetrator; it also aims at search for alternative, (positive) non-violent methods of conflict resolution; this includes *personal* conflict-resolution skills.

conflict resolution *(training)*: it raises awareness of individual responses to conflict and teaches ways of improving both communication skills and approaches to conflict.

conflict resolution *(paths)*: strength; negotiation, justice and personal transformation; of which the last proves difficult, unless conflicting parties are articulate enough to abandon emotion (irrationality) at least during the negotiation.

conflict types: depending on coverage and content: a) international; intra-nation; personal.

conflictual attitude: attitude maintained by an individual towards other members of the same nationality.

conflictual environment: where individuals are alienated for one reason or the other (viz on grounds of race, language or other real/imagined identities); attitude maintained by an individual towards other members of the same nationality.

contentious issue: sensitive issue, which, if left unresolved, could spark oft" conflict like river-water sharing between two riparian parties (polities).

crisis management: need proceed at three levels: international, national and personal.

critical theory: the starting point of critical theory (and in Burton's approach) is the problcmatisation of the origin of given framework of institutions and social relations. Like Burton's problem-solving approach, Critical Theory aims at clarifying possible alternatives for social change and suggesting ways how to transform the dominant social and political systems. And just like Burton, Max Horkheimer stresses that knowledge is not simply a reflection of a concrete historical situation, but has to be understood as a social force to generate social change.

critical theory: Burton's critical theory and 'problem-solving approach' are so inseparably conjointed as may be seen from his dictum: ". . . the solution is not the final end-product. It is in itself another set of relationships that contains its own set of problems . . . .problem- solving frequently requires *a new synthesis of knowledge* or techniques and a change in theoretical structure . . . The system of interactions is an *open one, i.e. the parts are subject not merely to interaction among themselves* but to interaction with a wider environment over which here can be no control" (John Burton: *Deviance, Terrorism, and War: The Process of Solving Unsolved Social and Political Problems:* Oxford: 1979).

culture of peace *(building a)*: institutional and collective effort of the society to eliminate *human aggressiveness* and, in its place, regenerate an attitude of' *caring'* for others, more like Gandhian lifeways; but this culture of peace can be best appreciated by knowing the other side of the coin (the anatomy of human destructiveness).

dialogue: interactive talk between (two) conflicting parties in a recorded (written) form, possibly being intently watched out by others; as distinct from a stand-off (or a conflict), it breaks the ice, presages hope, promises substance, and sets the map for road maps and 'change'.

diversity appreciation: education that promotes awareness of; and appreciation for, the many ways in which people differ and are similar in a society. It attempts to broaden definitions of one's own and others' identities.

dynamics of group work: operational advantage of team work to *internalise peaceable (social norms)* in the backdrop of decline of individual effort (personal power, pursuit & commitment) whose (academic) merits can be discussed in groups (for group commitment), but may fall flat on the ears of ex-hardcore insurgent, when its practical (dynamic) benefits can be proved by demonstration in group work in peace camp; also recall insurgents have hitherto lived and worked in groups away from families; (*see also 'alternative lifestyle' above*).

dysfunctional relations: that tense situation characterised by communication-gap between two (otherwise interlinked) social entities prior to the actual violence (conflict).

ethnic nationalism: an ideology, not necessarily chauvinistic, but usually does use chauvinism as a tool for dehumanising the 'other' and invalidating the latter's citizenship, territorial, economic or political rights. Ethnic nationalist chauvinism is more like racism *(also see 'racism 'for a difference.).*

education on bias. stereotyping: and prejudice: education on bias, stereotyping and prejudices raises awareness of chauvinistic attitudes and behaviour.

emancipation: 'True' needs can be communicated–and 'peace' can prevail - between different groups: 'self' and 'other' only under assumption of both 'adequate knowledge and un-distorted communication.' Burton's 'emancipation of suppressed group on par with privileged group' can prevail if the empowered group scales down its 'false' needs to 'true' needs, so that the 'needy' unprivileged group can have its unfulfilled needs satisfied. This according to John Burton embodies one and the only one 'problem-solving approach' for a conflictual polity.

emotion *(sulphur of)*: the suffocating smell and atmosphere of stand-off between two conflicting parties, in which only the identity brigands and votebank-seeking politicians find comfort, very much derogatory for peace

empowerment: power-'to'-people, rather than 'over' people; in one sense, it means people enabled to take more control of their lives and be less dependent on others; power-relationships are entrenched and exclusive of other individuals or groups, relating to gender issues, and marginalised and discriminated against groups (like SC/ Dalits & STs); K Boulding distinguishes between hard (coercive or threat) power and soft (exchange or integrative) power *(Three Faces of Power: Sage:* London: 1989) It is in the area of soft power, associated with bargaining, compromise, persuasion and transformative long-term problem-solving, which peacebuilders *utilize* to promote and create trustful relationship.

ephphata: an Aramaic word that means "to be opened"; as command of justices (of the subcontinent) seeking compulsive attention of the governing classes, the elites of both India and Pakistan, the two umbilical nations, to listen to possible settlement (conflicts resolution) as an urgent desideratum for both survival and development.

excluded group: people of a specific area which does not get (direct) benefit of a plan, scheme or project, as opposed to 'included groups'; hence the euphemistic: exclusive growth and inclusive growth (development) as emperor Ashoka of ancient India or most Manipuri kings; it could also mean 'desperate poor' or 'poor' sections of the people, who become undernourished, dejected, most forlorn, and hence people with 'injured' feeling towards the 'others'.

false consciousness: generally powerful groups promote false consciousness among the mass of the population; "as a result, less powerful groups aim at satisfying false needs as they' internalize need interpretations that work to their own disadvantage.' (Nancy Fraser: *Unruly Practices, Power, Discourses and Gender in Contemporary Social Theory:* Minneapolis: 1989) "Are you getting what you need, or just what someone thinks you need?" (William R Pottamchuk: *Govennnents' Responses to Human Needs*).

false needs: the corollary of Burton's 'Human Needs theory' is that there are both 'true' needs as well as 'false' needs. Needs are false if they escalate (work in the interest) of the powerful (dominant) group; or against the true interest of the working class.

game-changer: intervention mechanism whereby social interaction seem to change.

gender: of the three functions of gender: reproduction and production and social management, the third dimension is taken seriously only by the Human Needs theorists like John Burton. Gender is the social construction of social relations between men and women *on the one hand,* and as an *analytical category* to make "invisible" *gender-blind spots, categories* and *perspectives* in a (male-stream) society. These become "visible", once focused on gender 'spot on'.

'getting to *yes*': preparing (canvassing) such a questionnaire for (to be used during) peace-talks giving a straight-talk undeniably leading to the self-defeating character and end-result of conflict (including the trauma of refugees, children and women, annihilation/ destruction /devastation of property and social assets) by vivisecting and laying bare the 'anatomy of human destructiveness';

grey zone of peace: sort of peace characterized by full of uncertainties

hierarchical power structure: structural features of conflict;

historical education: a variety of formal and informal formats (forms) ranging from 'truth commissions' and books; to classes and historical workshops. It attempts to teach as objectively as possible, the events that influenced the war and the events of the war itself.

historical revisionism: graphic, one-sided history are taught to children as young as grade four, leaving no doubt who the 'enemy' is (T Walker: *Bosnian Schools Segregate Students: The Globe and Mail:* Toronto: Oct. 18, 1997.); alternatively, it also stands for skewed curricula, following the proverb: "Education is the battlefield."; scrambling to implant their vision of how an 'injured' community can become the 'imagined' community, given mass support in this highly politicized environs 'drama-in-education' *(for other methods see 'active learning methods'.)*

hostage to terror: surety (usually) held by a terrorist to warrant.

human rights education: that promoted knowledge, skills and attitudes consistent with human rights.

identity crisis *(childlike)*: this crisis of identity ensues when during a conflict any perception of self in the 'enemy' other (or of the other in the self) tends to threaten that gap and, much more like a child, one would have a regressive yearning; as a result, the 'enemy' is *'dehumanised'* in order to account, both, for the enemy's hostility; and also to justify the worst excesses of the in-group's behaviour towards that enemy; however both the in-group and the out-group are somehow alike in the way they behave towards each other;

inescapably preeminent is the fact that the *Self* and the *inhuman other* are similar. Further this continual threat to the Self drives a dynamic of enmity (Wedge B: *Psychology of the Self in Social Conflict: in International Conflict Resolution Theory and Practice:* Brighton: 1986)

intercommunity cooperation: term represent either an effort by educators to incorporate some form of inter-group cooperation into education programming when full mixing is not possible: or a cooperative project that has resulted from a mixed programme. Examples of the former are: civil society training projects which require some joint decision-making; examples of the latter are joint media projects initiated at mixed youth conferences *(see also intergroup mixing.)*

intergroup mixing: mixing participants from conflicting ethnic groups in one educational programme as a conscious policy designed to enhance learning.

intervention: (third party) intervention starts off by identifying the root causes of a particular conflict - the first step for any intervention

joint mechanism: post-conflict mechanism for intelligence-sharing, best practices and optimal outcomes so as to warrant optimal result to restore peace and normalcy through amity, trust, confidence among erstwhile conflicting parties; intelligence-sharing is the first line of defence in the 'terrorist terrain'.

media literacy: education designed to raise critical awareness of media, not only as an important part of a democracy, but also as text with its own biases.

mediation: an impartial third party having conflict resolution skills, and acting as a catalyst for prevention of conflict (positive education); but not an arbitrator (somebody who can impose his decision) in the mediation process; but who can transform the conflicting parties and teach them something about the 'other' person's perspective but also about their own; one who can deconstruct and then reconstruct an acceptable solution.

mediation *(process of)*: at the outset a time is set aside to explain their respective side of the conflict without interruption; to be followed by statement of their perception in the same uninterrupted fashion; next both parties are asked to explain the reasons - a focus on the *'needs'* of the parties, rather than on their stated *positions;* the parties are encouraged to *'deal with the problem and not the person'* — thereby ensure greater empathy between the parties by respectively understanding the viewpoints of the 'other' party; in fact, both then find that 'many problems are not based as much on *'truth'* as on *'perceptions'*; reciprocal appreciation of those divergent perceptions ultimately lead the parties to find some 'common areas of perception', which can further be widened through negotiations thus leading to a greater chance of *resolution* of the conflict.

mediator *(role of)*: a catalyst of violence; a provider of a context (medium or institution) in which both parties have the opportunity to speak and be heard by the other party/ies; but not an arbitrator (who can tilt the imbalance and impose his decision); may only suggest outcomes to the conflict for the benefit of both parties, whereby the parties continue to remain in control of the issues and remain responsible for, and therefore 'own', any outcomes; his mediation need however adhere to certain conventional process (*see process of mediation*).

mediation skills: non-partisan though and action, active listening, effective participation, empathy, creativity, and responsibility tor resolving the conflict by merely facilitating (enabling) the parties to resolve the conflict as their own problems, rather than a third party doing it on their behalf, or influencing/leveraging a decision from outside.

mediator *(peer)*: a pre-arranged (well-run) conflict resolution scheme, under which, a conflict is to be resolved by a peer mediator inside the scheme whose benefits accrue to the mediator and the parties which then reflects onto the school as a whole.

military solution: a policy stance or 'best option': in which an all-out effort is made to eradicate opposition through sheer strength of force, in which all resources such as war-hype, 'anger', ballistic missiles, artillery, and the best planners are marshalled to work up the entire polity and direct effort to maximum annihilation as in case of insurgency

national security: an inward-looking sense of feeling (secure) from ingress (invasion) of extrinsic forces.

non-conflictual attitude: attitude maintained by an individual towards some definite members of the same nationality.

pacification: a 'situation' where an unjust society is perpetuated through the suppression of violence.

participation: active involvement of beneficiaries (indicating a paradigm shift from patron-client relationship) in a meaningful way in the design of programmes, decision-making processes, project implementaion, control over resources, and the evaluation of programmes so that their needs and values are protected and enhanced, most relevantly to strategize peacebuilding.

peace: in a negative sense, 'peace as order' or zero violence-conflict; in a positive sense, an environment where people strive to transform society and communities into fairer and more just places to live. Not only is the 'direct' experience of violence tackled, but also the 'structural' elements in society which perpetuate as potential sources of conflict. Working for a positive peace means empowering people to become involved in non-violent change processes themselves, to help built sustainable conditions for peace and justice. Such a transformative approach to peace is not without its dangers because it means people will ask difficult questions about their social and political situation. It is within the more process oriented concept of positive peace where peace activists can locate peacebuilding.

peace *(inner)*: bagful 'insights of peace'; peace embedded in human mind in broad-spectrum and in-depth; kind of (meditative) bliss in one's innate peace-convergent attitude (soul), which is the obverse of the attitude and approach of peacemakers, peace activists, (secular) mendicants; (reminiscent of Buddhist monks' lifestyle who live by *parikrama* spreading gospel of peace and offerings during the pilgrimage; all out to assert forgiveness.

peace *(experimenting)*: peace activists out on a 'peace pilgrim' to experiment with other 'out lives' in transparent humility, seeking-forgiveness-for-intermediation' between conflicting parties, almost portraying the divine 'ambivalence of the sacred'.

peace *(positive)*: John Galtung (1975) differentiates positive peace from negative peace on the basis of narrow and broad approaches to peacebuilding. Further on, Galtung specifies that positive peace is closely linked to justice.

peace *(sustainable)*: transient peace, being maintained by concessions by two conflicting parties and characterized by (declared) willingness or intent for flexibility on key issues.

peace-hedging: act of hedging future against war, but ultimately peace-loving and seeking out opportunities for peace and trade; not because they arc nice (and enjoyable) but because they are accountable for (potential or past) losses.

peace constituency:

1. the contours (possible scenarios) if and when peace prevails, in particular stressing on all the prospects and positive aspects peace will help generate in the immediate and distant future; always across the wider canvas (spectrum), regions, nations, in fact the entire humanity;
2. another name of 'civil society', but one directed towards always inculcating or imbibing peaceful lifeways, in particular sensitive to the peaceful ways of problem-solving; about which Mahatma Gandhi once observed a little wishfully but almost as divine foreboding: "A day will come when the world will approach India in its quest for peace and India will become the light of the world."

peaceful tribe: a group of people with simple human needs (material and non-material) all self-sustainingly looked after and maintained at the level of the village itself: with little or no access (intervention) from external sources; e g sickness tackled by village medicine (cum-mystic) person.

peace activism *(promoting)*: to adapt to alternative lifestyle the conflicting parties (insurgents) (transforming into the other extreme of peaceful lifeways) activism need be promoted; given that they are alien to 'peace' as such, they are to be facilitated to "imagine (animate) peace' and to be shown 'inner dynamics of peace' — like the lifeways in an *'ashram'* (Gandhiji's hermitage) and a blow-up (21st century) module, to be used in 'post-conflict peace-building' follow-up effort;

peace education: education which raises awareness of the institution of war and processes that promote and sustain peace; conflict is an integral part of peace education; the animality in mankind is given to fight, but not skilled in *'peaceful problem-solving'*; but central in 'peace-education' is *'how to attack the problem and not the person'*; (UNICEF: Croatia Report)

peace study: a new interdisciplinary problem-oriented field: how to manage human conflict and bring creative development, rather than destruction resulting in (fair) justice to the satisfaction of conflicting parties; in particular the prevention of conflict: an ounce of prevention is better than a pound of relief'.

peace 'deficit' areas in conflict resolution: structural factors which render any unjust situation (inequality, economic growth and resource distribution) which stymie peacebuilding; political and institutional deficit (including institutionalizing a conflict resolution mechanism into the body politic); military-security deficit (and psychosocial dimensions of conflict and its pervasive impact in the aftermath through internalization of the peacebuilding process by afflicted populations (lay due stress on the psychological, spiritual and cultural determi-nants of violence).

peacebuilding: the practical aspect of implement peaceful social change through socio-economic reconstruction and development; rede-fined also as 'part of 'third-generation' peace approaches evident after the Cold War, when a reaction against simplistic approaches to building peace finally recognized the deep-rooted nature of conflict and its links to development. John Galtung argues that the issues of culture, human needs, and 'fault-lines of the human condition such as gender' have now enter the peace debate as crucial variables — to 'deepen' and 'broaden' peace before justice can be pursued. This does bring the peace process into tension with the very foundations of long-term reconciliation, viz 'truth,' 'justice; and ultimately 'forgiveness'.

peacebuilding from below: effort essentially directed to chart out sustainable and locally owned paths to peace and justice, as distinguished from championing 'third-party intervention.'

peacebuilding *(pillars of)* key to successful postwar peace building are: freedom of movement (release of detenus); absence of personal or group threats (removal of prohibitory orders); safe access to resources (income/employment avenues).

political education: education that raises awareness of political processes and the role of individual citizens in those processes. *(see also citizenship education);* it should reveal the way politicians try to influence or manipulate citizens and the ways in which social and economic structures influence and constrain political leaders.

political will: the conviction, dedication (mainspring of action) of political (government) leaders, as opposed to what they profess outwardly.

post-war peacebuilding model: "comprehensive efforts to identify and support structures which will tend to consolidate peace and advance a sense of confidence and well-being among people. Through agreements ending civil strife, these may include disarming the previously warring parties and the restoration of order, the custody and possible destruction of weapons, repatriating refugees, advisory and training support for security personnel, monitoring elections, advancing efforts to protect human rights, reforming or strengthening governmental institutions and promotion; formal and informal processes of political participation." (Boutros-Ghali; UN Secy-General. 1995)

potential nuclear terrorism: the possible stage (drama) of nuclear holocaust imminent once a jihad gets hold of nuclear explosives, as between two nuclear-neighbours like India and Pakistan.

problem-solving workshop: meet between heterogeneous elements, cultures, particularly representatives (delegates) of conflicting parties at the instance, and with the participation of: third (neutral) parties, consisting of active peace volunteers with media participating for widest dissemination of 'areas of disagreement' and possible scope for narrowing down the dissent through tension-reduction strategies;

psychodynamic theory: approach describing a "psychological gap" between 'good' self and 'bad' other, founded on a developmental projection of good and bad images of self onto suitable group targets (for more see identity crisis).

racism: an ideology used to achieve power and resources at the expense of a specifically targeted 'other.' *(see ethnic nationalist chauvin-ism.)*

reconciliation: the mode of bringing about normal peaceful relation between two conflicting parties, or effort thereto;

renegotiation of identity theory: the notion of the subjective self knowing the object other is replaced by a continuity between selves. Keller E.F. (*Reflections in Gender and Science*) proposes an alternative understanding of 'dynamic autonomy' suggesting that the fear of a loss of boundaries of the self is balanced by a fear of separation and alienation: i.e. a sense of self secure enough to tolerate both difference and continuity. "Epistemology (theory of knowledge) need be ontologically (existentially) grounded." In other words 'identities' are only meaningful in terms of interactions (which is their source). Individuals maintain conflictual attitude towards some members of a nationality, but non-conflictual attitudes towards other members of the same nationality and yet did not feel a crisis of identity because those identifications were negotiated in terms of these specific, interactive experiences with particular others.

restorative justice: the rehabilitation process of erstwhile dissenters/ insurgents onto' alternative lifestyles' before finally switching over to normal and peaceful lifeways; with separate arrangement for prison reform, where needed.

satyagraha: non-violent defense (struggle); Gandhiji's uniquely personal (articulately non-violent lifeways: including vegetarianism) approach to express his dissent in peaceable module; like' Salt Satyagraha', whereby import of British salt to colonial India was non-violently protested; India's freedom struggle was built around satyagraha; (cf resource nationalism in 21st century).

secular: religiously non-partisan; such a policy which conditions (irresistibly keep) public outlook adhered to it equidistant from all religions, and primarily to peace and stability, more than anything else (distanced from extremism in beliefs, outlook and action); for example the thoroughly Westernized barrister M A Jinnah, the founder (first) President of Pakistan — who had a bewitching fancy for his Malabar Hills residence in Mumbai (then Bombay) till his premature end within almost a year of Independence — is often cited to have given his first message to Pakistan being 'secular' in character and intent. Jinnah, it is cited, did not know how to read Quoran, which has become the constitution of even many a political leader..

social change: changes in the hierarchical power structure cause an environment of conflict relieving (empowering) the 'other', say the needy females, in a case, when females access more social power.

strategy of non-violent defense: the other (alternative) strategy to tackle a difficult (coercive) situation and to combat aggressive offence (and) to negotiate between conflicting parties.

structural inequalities: 'open' or 'hidden' power inequalities remain in any society ranging from: man *vs* needy women; privileged *vs* unprivileged; better-off vs. worse-off; etc. These are more often psycho-social, but most times they are in different templates (structural inequalities or hierarchical power structures), ultimately causing an 'environment of conflict'. Once a conflict environment remains (either open or hidden) a simple incident suffices to cause violence or a rift which can open up a wide conflagration. And unless timely controlled, intervened, or pacified, it might lead to a cycle of violence — more or less like a wild fire.

terror: an extreme form of violence, which perpetrated with the most annihilist intent to cause optimal damage and holocaust in the 'other' section.

terrorism: belief in terror as a policy weapon and bring about change in humanity.

terrorist terrain: the difficult but possible pathways (road-map) of a terrorist; often leading to a possible target with potentially maximum annihilation and traumatic stress on the public mind; with surprise as the most latent strategy of a terrorist (like that of a leopard) the latent approach of a terrorist at any possible target need be most scrupulously guarded against; in strict theory, the terrain is indefensible in case of a nuclear attack

trauma healing: trauma healing teaches about the impact of trauma and process of healing and grieving after a loss. Exercises may help people express their trauma or grief either verbally, physically or through art. Participants are helped to work through the grief process and to mourn.

track-two dialogue: strategic talk to broaden the 'peace constituency' at the track two (subsidiary) level as between experienced interlocu-tors of two conflicting parties, covering contours of conflicting issues for possible (strategic) nodes, as held in early 2010 at Bangkok between India and Pakistan; often before (or during suspense-filled inter-dialogues) peace talk.

working matrix for peacebuilding: the final plan to tackle first the crisis stage and issues for a period of 2-6 months; then the people and relationships for 1-2 years; then the institutions or sub-system for 5-10 years; and finally the vision of peace and desired future that all hope for and move towards and that will need work for generations.

xenophobia: the fear and dislike of strangers, especially those nearby and therefore in competition with the same resources. It is seen all over the world now. This often manifests itself in stereotyping of, and prejudice and sometimes violence against, the 'other.' While it can be used by ideologies, such as Nazi Fascism, it does not, in itself, constitute an ideology.

APPENDIX I

Conflict Mapping

Conflict is an off-equilibrium (abnormal) situation where in stead of peace violence prevails, and even the 'critical mass' of good and responsive citizenship is replaced by a tension build-up (varying from mere 'zero tolerance' order situation or 'negative peace' to extreme insecurity) in social spectrum including violation of human rights. Its recurrence has seemingly multiplied in direct proportion to the pace of social change, fueled by social awakening following greater awareness and the consequent mismatch between people's expectations from government on the one hand and increasingly poor delivery of governance.

To ascertain the multi-factorial and dynamic process of conflict, academia started research on structure and context of complex conflicts from both subjective and objective perspectives; if the particular social-psychological perspective on conflict "is as informative and constructive as is assumed, or whether alternative perspectives might have as much or more to offer"; but such mapping is highly subjective (transitory) and its transcendence over time after the conflict situation is negotiated; its transient nature is also because it is not 'objectified by the knower'; in other words, far from existing in a fixed form, waiting to be observed or known, the (social) reality is constructed and negotiated intersubjectively.

Psycho-socially therefore another possible explanation of the identity crisis, strongly suggesting that identities are only meaningful in terms of interaction that is their source; same individual may maintain conflictual attitude towards some members of a group, and non-conflictual attitude towards other members of the same group; hence did not feel a crisis of identity; and this was because those identifications were negotiated in terms of these specific, interactive experiences with particular 'others'; but once the quality of those interactions changed, or where experience of non-conflictual interaction was denied to the individual, the identification of Self and Other was re-negotiated; using such a roadmap it becomes possible to systematically identity key historical and contextual influences on development of the conflict; to identify the parties; list the issues; explore the conflict dynamics (emergence, polarisation, escalation, psychological factors such as stereotyping and prejudice); analyse in terms of possible conflict resolution approaches; decide whether to intervene (accessibility, timing, alteratives, probability of success); consider concerns of neutrality and credibility; and check available organizational and individual peacemaking skills (including trust commanded from the stance of both parties);

An important element is a careful analysis of the situation that should include identifying particular interests and the power constel-lations of different identity groups, in particular ethnicity background, ethnic alignments (re-alignments). To start with, conflicts need be categorised into four broad levels, although in most cases these categories are found interlinked:

a) *Intrapersonal:* how we as individuals understand its root causes deal with; and of course also interlinked in successive stages overtime;

b) *Micro-level:* within (inter-personal) and between 'communities' (villages, ethnic groups, leaders and people, neighbours etc.) institutional conflict;

c) *Meso-level:* conflict involving militant groups, local government, international NGOs and local NGOs (including NGO-rivalry), UN agencies, national coalitions;

d) *Macro-level:* violent conflict at the national or international level;

Whilst peacebuilding programmes need be defined specifically for each level of analysis, it would be useful to develop contingent and complementary frameworks or schemes incorporating the above level for attention of specific-level peace activists. In complex conflicts (usually with many conflicting parties as in case of Manipur), this requires sustained cooperation and coordination between the various agencies. It is also important to consider the broader moral and ethical implications of such interventions (Lewer and Ramsbotham: 1993). Needless to add, a historical, political, economic and social origins of violent conflicts–

which often lead to dysfunctional relations between groups and individuals in a country–should be borne in mind while considering peace building activities.

Equally important is a tine-grained analysis of the local 'communities' or 'constituencies' in which such interventions are planned. One would assume that this is the prerequisite of a good (textbook) development planning. But household surveys such as Participatory Rural Appraisal (PRA) surveys and other technical research should include conflict analysis, householder's perception for conflict analysis (mapping) and a mechanistic understanding of the socio-cultural relationship and fabric of the communities–broken up in terms of the conflicting and non-conflicting parties.

By way of conflict intervention module, it is also important that the third party or NGO taking up the role of mediator, arbitrator, advocate, trainer, witness, supporter, counsellar, therapist etc can tackle both relational and structural elements of a conflict: Examples are given in the following box:

a) *Intra-personal:* gaining personal knowledge of causes, general theories, concepts and approaches to understanding peace and conflict; personal change and transformation workshops;

b) *Micro:* non-violent conflict resolution training; 'human rights awareness-workshops'; exchange visits and amity camps; cross-cutting programmes, trust-building and conflict-mitigation programmes;

c) *Meso:* national organisations and coalitions for peace and reconciliation including business, political and religious coalitions; systematic and national initiatives in *peace education,* development programmes, media monitoring;

d) *Macro:* offers of mediation from the international community; attempts at persuading government and opposition parties to have a united and coherent strategy to the conflict, and not to use it as a political football (an example is how Conservative and Labour parties in UK deal with the Northern Ireland/IRA conflict); involvement of peacekeeping and peace enforcement forces

APPENDIX II

Limitations of Rational Human Expectations

He sees the future not as an arrow he shoots ahead of him but as an arrow that arrives at his feet. She tells an anecdote about a friend who liked to joke that he was a Bengali trapped in the skin of a European. Her own sensation is not so much of a terrible mistake, but of discovery and arrival; a kind of "fatedness" Not that she claims to understand anything. On the contrary, she believes in being willing not to understand, while simultaneously remaining curious and open.

Against most rational expectations, almost all marriages between ethnic groups have worked all along. These are lessons *in* the limitations of rational expectations. Likewise, *Come on Shore and We Will Kill and Eat You All: A New Zealand story,* is really about mutual incomprehension, illuminated, if not dispelled by its author's (Christina Thompson) romance with the Maori, in particular one, called Seven, who she married.

Christina Thompson is a New Englander from a trim town outside Boston with a white church and a green. Seven belongs to Ngapuhi tribe and his family lives in a ramshackle settlement at the end of a dirt road. Ms Thompson is, above all, an intellectual *in* the tradition of the Enlightenment, an editor of the academic *Harvard Review.* Seven with his belief in ghosts and aliens, is the very man that tradition hopes to enlighten. She weighs options and makes plans. He sees the future not as an arrow he shoots ahead of him but as an arrow that arrives at his feet.

It all falls neatly into Western categories about "advanced" and "primitive" peoples. But Ms Thompson is not so pat (ready) as that. She tells an anecdote about a friend who liked to joke that he was a Bengali trapped in the skin of a European. Her own sensation is not so much of a terrible mistake, but of discovery and arrival; a kind of "fatedness." Not that she claims to understand anything. On the contrary, she believes in being willing not to understand, while simultaneously remaining curious and open.

The initial contact especially intrigues her, the bafflement on both sides, an equality of wonder which, however briefly, belies the more common tale of ferocity or exploitation. She describes an evocative episode in 1642, when a Dutch ship is greeted by canoes off New Zealand, and the two sides, after fruitlessly calling to each other in their own languages, resort to music–a series of trumpet calls, first one and then the other, until darkness falls. Sadly there was death in the morning, but as cultural exchanges go, it did not start off too badly.

Then there is the story of Charlotte Badger, a convict who was one of the white women to land in New Zealand, in 1806. She lived among the Maori and twice declined to be "rescued" by ships sailing for England. Ms Thompson recounts the little that is known about her and, drawing on her own experience, imagines, beyond the perils and discomforts, the attraction that must have existed on both sides, in particular the discretion and forbearance of the Maori whom she herself encountered in her husband's village. Unlike Charlotte, Ms Thompson returned home, anxiously bringing Seven and, by then two sons, to her parents' house. But the expected clash never comes. There are tricky moments–for example when Seven, an ingenious mechanic, fills the drive with dumped machine parts. But otherwise, he fits into their lives like the natural gentleman she has read about so skeptically in "noble savage" literature. All through the book, the author makes a *point of wrong-footing herself.* Predicting, interpreting and being clever do not seem to work.

It is Seven's brother who points to *the hidden story i*n Ms Thompson's enterprise. Shortly before leaving for Boston, she tells him that she intends to write his family story; "Write your own first", he says. It was *a shrewd retort.* Her own ancestors, Englishmen with American-Indian blood on their hands, turn out *to be not so different from the people who have impoverished Seven's ancestors.* History, in other words, brings her closer to Seven than she could have imagined. It is the charm of the Thompson story that the circle never quite closes that in the final pages *(read* metamorphosis/evolution stage), history itself becomes another foreign country, another adventure." (*Economist,* Aug 9, 2008: p. 78)

APPENDIX III

The Security Angle

Having all through the last 4 decades involved in counter-insurgency operations (CIO) in Manipur, the Central Security forces must have had enough bottomside perspective of the conflict situation from the overall mainstream stance. Those who saw action here have even risen in the army ranks to the top echelon. By their own admission, it was not all through a success story. Yet having achieved sufficient ground so far, they seem to speak in terms of a possible "turn-around" in the entire tangle, for which they seem bent upon testing the ground by feeling the pulses of local elite. Here is a summary of such a recent interesting interaction right from the horse's mouth: Public Relations unit: Ministry of Defence: The Red Shield Division held a one-day seminar on Conflict Resolution in Manipur at Leimakhong today. The General Officer Commanding 3 Corps Lt Gen NK Singh, AVSM, VSM, also attended the seminar. The purpose of this seminar was to explore the prospect of peace by charring diverse opinions and view points on the issues afflicting the State. An important highlight of the seminar was that all participants belonged to the state itself giving the audience an insider's perspective on the various contours of the conflict and violence. The themes of the seminar included 'Understanding Ethnicity and Alienation', 'Role of various Institutions in the Peace Process' and 'An Approach to Conflict Resolution'. The seminar was attended by people from all districts of Manipur and included representatives from various schools, colleges, tribal and student bodies, human rights organizations and the media. The keynote address was delivered by Justice (retd) MA Shishak. The opening address was delivered by Maj Gen D S Hooda, VSM** .

"3. I think this is a great opportunity for all of us to meet and exchange our views, ideas and perceptions on the problems beseeching this exceedingly vibrant State. I use these words carefully. There is a vibrancy which you seldom see anywhere else - in the people and their songs, the dances, the football match in every locality. But there are also problems which are preventing this State from achieving its full potential. Today, we have a distinguished panel of speakers representing all walks of life in Manipur. We also have a representative audience. Since all of them belong to the State, we hope to get an 'insider's perspective' on the opportunities that may lead to a resolution of conflict in Manipur."

"4. Although a majority of the NE region is beset with conflict, I believe that the conflict in Manipur is not only unique but its resolution is of paramount importance for all of us.

(a) Manipur inside India is mini India. It is a melting pot of diverse cultures, religions, castes, languages and ethnicities. Despite these diversities, the fact of unalterable geographical reality, a history of common past and a shared vision of a brighter future should bind the society together, both in India and Manipur. I think Manipur can become an example for the rest of the country to follow or conversely it needs to follow some of lessons in nation building that India has learnt since its Independence. However the issue of managing plurality remains important, both for the nation and the State.

(b) In Manipur we are experimenting with Suspension of Operations (SoO) with some underground groups, much after a Ceasefire agreement was signed with NSCN in Nagaland. Such agreements have reduced the level of violence, but the absence of violence does not mean 'peace'. Conflict resolution cannot be viewed in terms of peace accords and Ceasefire or SoO only. There is a need to explore the larger dimension of peace, which is inclusive in nature, long lasting and integrates those who abjure violence into the mainstream of society.

(c) The militarization of civil society and fear of the gun in Manipur is a cause of major concern. When the common man feels insecure, all of us must sit up and take notice. Such an environment will only embolden the underground and preclude even the initiation of dialogue which is the first step to conflict resolution.

(d) One of the primary reasons for spawning of insurgencies in the N E region was a feeling of neglect and alienation, It is said that the ripple effects of conflict in the NE reverberate within the region itself' as the region is geographically bottled up. Post liberalisation, our nation is on the threshold of a major economic development — but the GDP growth wave, sweeping over the rest of the country has not been able to penetrate this geographical region."

(e) However, India's' Look East Policy' (LEP) has the potential to break the so-called geographical isolation. Since trade and economic cooperation with the countries of South east and East Asia will increase in times to come, and Manipur lies on the road to this part of the world, it is imperative that we create requisite infrastructure here. This will be a 'win-win situation' for the people of Manipur. India's LEP cannot succeed unless the nation looks at the problems of NE states more closely. The success of LEP is important for India's committed economic growth and is inexorably linked with violence in the NE region in general and Manipur in particular".

"6. This seminar on 'Conflict Resolution in Manipur' has been organized to have a free exchange of ideas/ views. I would request you all to give a patient hearing to all the speakers you may not agree with their view but as Robert Frost said, "Education is the ability to listen without losing your temper or your self confidence."

The closing address of Lt Gen NK Singh was equally striking:

"Corruption and extortion are endemic and seem to have become a way of life. There are multiple interest groups who are propagating a brand of divisive politics which has given rise to multiple perceptions. The challenge of Manipur is perhaps to manage perceptions that construct everyday life". He further said, "hence people must make use of RTI Act to check institutional wrong doings at various levels".

Lt Gen NK Singh AVSM,VSM
GOC - 3 Corps

APPENDIX IV

The Third Dimension to Nagalim

Territorial integrity of Manipur: RS. Jassal interviewed by Dominic Maring, a research scholar:

1. **Why don't the Meiteis allow Nagas to form 'Greater Nagaland' ? What are the reasons?**

   R.S. Jassal: "Oversight. It's quite obvious 'greater Nagaland' means clipping of territories of Nagas reducing balance of Manipur to the size of one dist in UP /Bihar affecting century-old socio-cultural fabric of hills and valley. Map of Nagaland circulated by NSCN (IM) is a pointer to their political agenda which can be termed as second stage of amalgamation of territories. And as per intelligence agencies, and common sense, stage three may be adding of areas under occupation by Myanmar into Nagalim. Thus based on territorial, cultural, administrative and military strength (presently both NSCN factions & NNC) claiming as a right of creating a separate nation for indigenous people may also be influencing Meitei vision." Who knows one may also prompt Kuki-Chin groups too to look Mizoram, then imagine the left over size.

2. **On 18th June uprising 2001, why did the Meiteis make such a violent protest?**

   R.S. J. Because of savoury and unanimous statements by the Ms LA, Ms P, state BJP/Congress hierarchy, all assuring of not losing even an inch of Manipur territory — despite con1radictory announcement of extension of Ceasefire (CF) to all Naga-inhabited areas by the Prime Minister (Shri AB Bajpayee) — people from valley resorted to violence under surcharged emotions against Central leadership but unfortunately got hell-directed against the State legislators, Assembly precinct, CM's office and Ministers' bungalows including Raj Bhavan under seize.

3. **Do the Nagas feel unsafe to live in Imphal valley? If so, what do you think? And what are the reasons for it?**

   R.S. J. Not entirely true. But, yes, in the event of Nagalim coming up into form, properties and lives of Nagas in and around Imphal may get damaged hence fear psychosis exists. Reasons are not far to seek. This happens in any ethnic (ignited) conflict situation.

4. **Why is the conflict between Meitei and Naga happening?**
R.S. J. Nagas think reasons for economic neglect into Meitei supremacy since centuries, initially due to autocratic authority of Manipur kings, but the State Assembly. Constitutional protection to Nagas in the field of employment, educational scholarships, acquisition of properties in the valley and denial of similar privileges to Meiteis is an other source of irritation. Originally people from the same homogeneous stock got divided into Hindus (from Sanamahi faith) and Christians (from animist Pagans); but turned heterogeneous by changed faith and design. Also ill treatment of hill people by Meitei kings in drills of subjugation and unbearable primitive methodes used in tax collection of the past and acts so related ingrained the psychic reflexes of superiority and inferiority on either side. Rebels or insurgents! they change their banditry. But we have to negotiate with them, as we don't have the means to fight them. Do you have any better idea? Ethnic tension arose much earlier because: (1) extremists each side paddle the dangerous myth there is a zero-sum game between the races; (2) That the way to win it is to take to the guns; (3) Politicians also risk helping the extremists on each side in their bid to gain vote bank. It thus raises a valid question: "Is democracy ever beneficial for all occasions?" So Ngas might be feeling joining main stream of Naga available next door.

5. **(Supplementary Question) Can you quote some authentic examples?**
R.S. J. Yes, L. W. Shakespeare indicates in his book *History of Assam Rifles* (p.26) that "Togwema" of some clans of Marams presently called Willong, one-day-long-march from Maram at the mouth of Zupvu valley (meaning by Zuko valley) was a one-time heavily populated village indicated by the most remarkable Stone-henge of gigantic monoliths which would not be possible for a population as of then in AD 1911 or 1923 (seen by Shakespeare in 1900, T.C. Hudson in 1911 and Hutton in 1923). It was definitely

the contribution of a much larger population in the far-off past in order to procure and drag stones into position, probably from a considerable distance over steep inclines. He lamented the village being so small as of then. For sending 2000 persons in Labour Corps to France during W. W.I on demand from the British authorities, tribals were detailed. Hinduisation of the valley and Christianization of the hills drifted their two peoples further apart quite subconsciously but dangerously. Muslims, while Gandhi of ....

6. **When did the conflict begin and why?**

R.S.J. Naga Club of the then (1929) Naga Hills district fanned their cultural activities into the neighbouring hills of Manipur too gradually, while Manipur Pan Club of 1939 activated cultural activities to the then Gauhati base. Both somehow drifted into influence (tendentious ideology) of communist leaders: AZ. Phizo and Hijam Irabot (though at a much later stage). Both met in Burma during the fifties, Both failed to reach China through Thakin Ahan Tun, the then chide of Communist Party Bureau (CPB) of Burma. Both failed to bring revolution through communism and resiled into Socialist modes. Both thus confined their activities to Cultural Revolution for people under their areas of influence but political events moved so fast: British were leaving Indian subcontinent in a hurry, forced by post-World War II compulsions in the process dividing India into two giant political blocks: Congress and Muslim League. Jinnah became father of Islam and Gandhi of Indian. If one became Muslim-dominated, naturally those on the other side of the divide became Hindu-dominated. Sardar Patel meanwhile engaged himself in consolidating India's new boundaries with NE region as a unit, while Pandit Nehru was busy tackling Chinese premier Chou-en-lai, Nasser, Tito of the then non-aligned nations of the two-superpower led world blocs. This has its own retrospective effect on both Nagas and Meiteis. Meiteis did not take issue initially, since Nagaism in Manipur was conspicuous by its absence. Meanwhile Nagas became apprehensive of their entity extinctions under the Indian rulers as they were neither Hindus nor Muslims–

if not flabbergasted by the fast-moving territorial consolidation of Sardar Patel. The origin of the problem was psychedelic in nature. Meiteis derived false satisfaction taking Naga integration as a granted fact (with centuries behind). But gradually they became bewildered, whether, to look to main leadership from the then Naga hills for their future or go to Meitei leadership at Imphal and remain under Hindu-dominated leadership as the majority of the learned and enlightened Nagas had become disillusioned with Meitei leadership for reasons all inverse and insiduous.

7. **Where did the issue originate?**
   R.S.J. Ethnic identification, speedily lapsed into ethnicity construction, can be termed as sprouting of the seeds of social separation between the hill tribes and Meiteis. Hill tribes who were till a few decades ago not conscious of being called Nagas, and were happy called as Tangkhuls, Paomais, Marams, Chirus, Thangals etc. welcomed cultural initiatives from main Nagaland. Since there is no provision in the Indian Constitution to check such developments, it has subconsciously taken shape of territorial integration for Meiteis to retain and conversely for Nagas to get under one administrative unit of the main Nagaland to achieve due contiguity and triobalism, despite the fact of Meiteilon being the lingua franca of all hill tribes and tribe-to-tribe lingua franca of Manipur. Nagas in Nagaland tribe differ in dialects from say Lothas to Yiumchungers, Khiamungans, Konyaks, Tikkirs, Semas and others. This may be perceived as the origin of the issue i.e. territorial integrity -v- one Naga administrative unit but they have Nagamese as a common link dialect.

8. **Who is more involved in creating the conflict?**
   R.S.J. (a) youth (students' unions);(b) Meira Paibis/clubs; (c) Politicians (d) Underground groups; (e) Pressure groups (UCM, AMESCO, AMUCO, UNC etc.) (f) Pressure groups under direction of UG groups & some UG intellectuals.

9. **What are the dominant issues people are discussing?**
   R.S.J. <u>Dominant issue.</u> It's about dilution of integrity of Manipur territories for the Meiteis, because if Naga integration takes place,

CCpur, Tengnoupal, and Moreh may also fall in for Mizoram or other contiguous areas of mutual interest. For Nagas, it is, a waste of effort of the last 50 years if they are unable to integrate into Nagaland. Second, they apprehend neglect of development of their areas since Meiteis do not agree to extension of Sixth Schedule to hills. Kukis feel peeved off in between for non-creation of Sadar Hills as a revenue district, whose contribution and historical links to Manipur's socio-cultural and political economy has been unequivocal.

10. **Who are the beneficiaries and who are the victims? What kind of treatment has the community experienced? *(Include gender analysis when identifying beneficiaries and victims.)***
    R.S.J. Victims are minority tribes, miniscule ethnic groups including Mayangs & Gorkhas. Beneficiaries are the UG organizations and the frontal organizations. Big business houses are also harvesting fortune by swindling with market management as everything is ferried in from the mainland. Supply of spurious medicines discarded (refabricated/repaired) machinery and adulterated food items etc. which find lead routed Manipur to corrupt practices, turning Manipur into a melting pot of disgust and despair.

11. **Who is economically or politically benefitting from the conflict?**
    R.S.J. UG organizations and black marketeers. Powerful going more powerful.

12. **Who are the decision-makers of the various parties/group? *(Use gender perspective* in *answering)***
    R.S.J. Central leadership of each UG group and select intellectuals acting as frontal leaders.

13. **How has the community faced or coped with similar problems in the past?**
    R.S.J. Past Manipur never reached this pitch *(read* tipping point). Political leaders had displayed strength in political acumen and vision. Now most influential leaders are marginalized on one ground or the other due to polarized UG leadership and corruption ethos

reverberating (moving stealthily) under the shadow of the guns. Manipuri kings were really different souls or god-kings: differently moulded as divine Pakhangba.

14. What are the roles of the Government and non-Government institutions in the conflict?

R.S.J. That of understanding, toleration, conciliation, persuation and mediation to go upbringing new generations with sound morals and ethics to become sound pillars of new social order.

15. Are there any obvious difference: ethnic, religion, and socioeconomic between the parties involved?

R.S.J. Religion and ethnicities do not appear to affect much. But SC/ST status for tribal, unequal land laws, varying ownership status of land and settlement, typical forms of political constituencies forced on the hills (being large in area but less in population) have created strange situation i.e. even after statehood achieved (1972) no improvement in their political strength of representation ratio of 40 valley vs 20 hill MsLA strength fixed in taking any decision. Economically valley is much more advanced and rich compared to hills, still lacking in education, communication, healthcare. The UG interference makes the situation demonstrative as government machinery fails to provide clear and fair governance.

16. Are there any activities which unite inter-faith leaders and social classes?

R.S.J. Intermarriages and age-old social ties lingua franca, common trade, inter-dependencies on hill and valley resources, and above all, corruption unites all. Manipur's connection with outside world actua1lyweakened for want of commensurate infrastructure (colonial, vestige) so the engine for innovation became much less powerful–failed to sustain the multi-faceted needs of its plural ethnic groups.

17. What are the claims of the parties: Meiteis and Nagas in conflict?

R.S.J. Meiteis claim that Rajas/Maharajas ruled over hills by heredity claim hence they were monarchs. Nagas claim they were never ruled by Meiteis and discard Meitei's claims to assimilate hill areas through conquests. Rule of valley kings extended only to valley. Nagas were ruled by the Vice-President of Durbar direct who was always a British which denied Maharaja much say in the hills. It is not only substantiated by history, but is also irony of fate: i.e. administration of [people by double standards. More than Meiteis and Nagas, other contenders are there to the conflict and in the resulting peace talks. In fact the talk itself becomes no longer a two-sided chess. It becomes a "36-dimensional chess", if not now in serial talks to be had with as many ethnic groups. Already Nagas had a premonition, that's why they are trying to make the current talk the "end-game" a two-sided talk with only GOI. If any compromise is reached, the 'multi-dimensions" in the "end-game" of the chessboard will include. You can then no longer just say: "Go to it, Chidambaram. Go to it Muivah. Just get it done". You must address all 36 groups and say "Just get it done!"

18. What are the perspectives (objectives) of the conflict from all sides?

R.S.J. UG Naga's vision is Nagalim and UG Meitei vision is revival of administration prior to British interference. You may call it independence or abstract disobedience to Indian democratic system of governance as the UG outfits from valley and majority of political pundits of both valley and hills value on power hold i.e. political strength which means elections; and election breeds corruption; and hill MLAs numbering 20 against 40 (valley) will always be at a disadvantage and presently there is no way out to solve the population ratio disparity. Every culture defines for itself the perfect point between 'squalid' and 'over-fastidious'. [Katherine Ashenburg: *The Dirt on Clean]*

19. Is there any indigenous, traditional or local way to address/ solve the conflict?

R.S.J. All these appear to be one and the same thing. Yes, through cultural track i.e. by composing songs on unity and respect for each other like long back. "Ougri" was sung by Meiteis in Mao area to strike the chord of unity and sense of forgiveness and not to revolt against Maharaja Chandrakirti Singh was successful. Bhupen Hazarika had also sung a song of love and forgiveness to Nagas at Kohima in 1964 in the midst of the bloodiest bloodshed, which became the forerunner of the Ceasefire then, wearing tribal cloth and Meitei cloth at least once a fortnight. Organise clubs and unions area-wise, community-wise and so on. By all means disseminate Obama's message: "We can no longer afford to build ourselves by tearing each other down."

20. Is there any local symbol, used by the community for certain events which describe conflict or peace?

R.S.J. No. Except white flag by Naga and black flags by Meiteis during 2001-02 on extension of Cease fire (CF) to all Naga-inhabited areas and its roll-back.

21. What are the symbols of customs and what are the meanings?

R.S.J. Kanchenkui–a typically designed crossed wooden piece as decorative item on each building of importance in valley and hills alike; like welcome gates and so on; The church building at Senapati is adorned with same; so also many other structure in hill and valley.

22. How do people benefit from existing natural and economical resources common to both?

R.S.J. This has brought the Central government on tenterhooks. Under cover of disturbed conditions the state is being divested of natural resources: timber etc. No development work is being allowed to progress without coerced cuts for which discredit goes to the State government. People with muscle and money power indulge in illegal trading organizations funded by even foreign money, violate FERA regulations and State is made to suffer economically. If meaningful planning is done people transcend district boundaries and adopt policy of share and care, economically Manipur will not only become self- sufficient but has potentials for earning foreign exchange: e.g. though power, timber, horticulture, tourism and sports. It has tremendous forest potentials and is rich in natural resources.

23. What is the relationship between individuals, groups and communities in conflict areas (particularly in Imphal area)?

R.S.J. A peculiar relationship of firm faith in oneness with riders of suspicions, doubts, apprehensions and misapprehensions exist. Passions and emotions have blurred the vision on both sides. But remember: history repeats itself. This revision is in the form of a well-balanced, mature, and timely call with studied considerations of pros and cons like that of Hokishe Sema to effect a change in the mindset of peoples. That is the status held by many Naga intellectuals like Niketu Iralu, S.C. Jamir. Likewise, many senior citizens like K. Prongo, Lanu Toy (ex-CE), Tochi Hanso of Nagaland, R.K .Theko, R.V. Mingthing, Ngaite IAS (retd), Rishang Keishing, Thangkhanpao also enjoy. Same needs to encash Meitei leadership which unfortunately is not one person but proliferated. Every leader moves in his/her own realm. Credit must be given to IM and church leadership or narrowing down proliferations at intra and inter.1evel among various Naga UG organizations to oneness, by coercing them to think collectively. Meiteis need a conclave of thinkers, social activists, religious leaders, and intellectuals who can issue guidelines for social reformation for unity and conciliation among themselves and guide Meiteis and Nagas to chart a peaceful road-map by filling potholes of distrust.

24. How do the people deal with management of their land in the conflict areas? Who benefits more from this style of management?

R.S.J. Hill people can own property in the valley. But valley people cannot own property in hills. This is the biggest eyesore. So on unfortunate and opportune conflict situation like Naga-Kuki clash (1992-93)–apparently for ethnic cleansing at the initiative of NSCN (IM). Both groups suffered loss of houses and property–Kukis in Naga-dominated and Nagas in Kuki-dominated areas. Of 22,327 sq km, only 1600 sq km is forest-covered under different denominations (including village forest)–mostly in hills. In fact, the forest coverage shows 16% surplus over the national forest policy targeted coverage. The entire hill shows a grave pattern of underutilization of land as most hill people still continue to practice traditional jhuming. Hence jhumias make a living out of illegal taxes on extraction and transportation of timber, while state forest department earns just a pittance. Security forces were often reported carting away good quantities by stealth, evading revenue payment to the State in spite of Supreme Court ban on deforestation.

25. What are those policies and regulations of the Government that might have caused a conflict?

R.S.J. Land laws granting unequal rights to own, use land between hill and valley people; non-issue of patta to individual in hills; non-extension of Sixth Schedule of the Indian Constitution; non-implementation of Sadar Hills as a separate revenue district; Imagine entire Thoubal valley district people; as also, the Jiribam subdivision (located at Manipur's border close to Assam's Cachar district:) OBC, non-ST people asked to vote under Outer Manipur Parliamentary Constituency. Such miss-match of political rights to vote or contest election even after six decades of Independent India speaks of gross non-understanding the ground realities typical to Manipur.

26. What is the role of the government in conflict- management?

R.S.J. Government–including politicians, bureaucrats, SF need help NGOs to render positive peace education and to dispel any notion of negative peace by inculcating various programmes and people-to-people contact programmes at the inter-group level/. Likewise ISKCON and various religious groups and missions must join hands in bringing back amity and cohesive spirit among people at the grassroots level. Meanwhile, people's outstanding political and other genuine grievances need be redressed, while Planning Commission, NEC, etc need emphasize on such programmes to narrow down regional and intra-regional socio-economic inequalities. In fine, Dr Radhakrishnan's message need be recalled: "The way of peace requires that men and nations should recognize their common humanity and use weapons of integrity, reasons, patience, understanding and love." Corruption has to be curbed down with ruthless measures. State government must display fair and dare governance. Operating of trades, commerce and communication with Burma & Bangladesh with open border agreements will make Manipur a very viable and strong unit of the Indian Union. With Tipaimukh Hydro Project and Railway link (trans Asia), Manipur will have a new FACE.

Thank You.

(This interview was widely published in almost all the local English Dailies in Manipur and North East Herald, Nagaland)

APPENDIX V

Proclamation of his Highness Maharaja Bodhchandra Singh on the Inaugural Function of the First Manipur State Assembly on October 18, 1948

Introduction

This is the first time, I, as a Representative of Manipur Crown and Monarchy, unbroken and safeguarded with the blood of my forefathers for about two thousand years, address the first Assembly of the Manipur State on this historic day following the modern practice and British style. I consider myself fortunate to see the birth of an Assembly in the State during my time and again I feel proud to be able to place on record and make it known to the rest of India that though painted otherwise abroad, my State for the first time in India has, just as the sun rises earlier in the east, taken the lead in the direction of democratic government, based on adult franchise and joint electorate in the line of which India is soon following. I take more pride in that though in countries like England, responsible government has been the outcome of the blood of so many persons, there has been in my State no bloodshed since I took the initiative here.

The New Constitution

While I have been all along anxious to accord responsible government, I have been sincerely watching as to the practicability or impracticability of the Constitution Act. I now bring to the mind of the people that I had transferred my powers and responsibilities other than those of a, Constitutional Ruler to the State Council since 1st July, 1947 before the lapse, of the British paramountcy and since then, I have already remained as a Constitutional Ruler. But most of the people innocent as they are; have 'been imputing to me the actions of the Council. My only earnest wish is to see a more prosperous Manipur. Though the literacy is below five per cent, you have had the best opportunities, as afforded by a responsible government to serve your country to the fullest extent you desire with toleration and concord to make Manipur, a greater Manipur.

Aim

Administrators and citizens of the State, in order that ,the integrity and separate entity of the State is preserved for ever .as heretofore ensured by Pandit Jawaharal Nehru, I wish to attach the highest measure of importance to detailing some account of what the State Manipur is –I mean the State and people, whom you now represent and on whose better administration you are now out to embark.

Historical Background

To begin with, I feel a sudden emotion in my heart–I mean the sweet reminiscences of the past history of Manipur. Let the patriots fight for today. I will see that yesterday shines once more before the people, ignorant of the beauty of yesterday. I am sure, a look back to the history of our land will at once inspire your mind towards greater and nobler deeds to prove your worth as worthy sons of worthy fathers.

Lo! the sweet and green valley with the sweeter and greener hills all around along the easternmost part of India–a smiling land, allowed by footprints of heroes and admired as the Switzerland of India by strangers. It is Manipur, a salubrious land of hoary antiquity–a land of gems. Geographically, it is singularly isolated from the rest of India. Strategically, nature impresses on her a stamp of unity by fully encircling her with lofty mountains and hills making it an impregnable fortress and bulwark of defence doing the sentinel for the millions in India. She was a star in the eastern political firmament of India. Strong nations of the world covet her natural position. Such is Manipur.

Read her political history from A.D. 24 She had her dominion over a wide area extending as far as the southern portion of China in the north, the gold mines in the Sibasagar valley, the river Chindwin in the east and south, and Chandrapore (Cachar) in the west. Her present area is 8,650 square miles plus 7,000 square miles of the Kabaw valley, including 7,900 square miles of the hills. Read the war operations of Maharajah Garibaniwaza in 1711, which shows how far the spirit of heroism and independence ran rampant in the veins of our forefather. Read the Treaty of Maharaja Gourashyam and Mr. Verelst of the East

India Company at Chittagong on the 14th September, 1762 which shows the vast wealth and economic status of Manipur. Read the Treaty of Yandaboo, dated the 9th January, 1834 ratified between Maharaja Gambhir Singh and the Company. This treaty left for us a good heritage, namely the Kabaw valley of 7,000 square miles for which we now get from Burma Rs. 6,270 as annual tribute which according to the spirit of the terms of the Agreement, will cease when the area is reverted to our State. Maharaja Gambhir Singh died of disappointment for this loss of territory under duress. Sreejut Nara Singh's role in redeeming Manipur from Burma, when she was devastated and dismembered for 7 years was very great. Manipur was thus, by reason of her vast territories and econo-mic resources, self-supporting as a self-contained unit. Her sphere of influence and mercantile activities was very wide. During the British rule, foreign goods gave a dulling effect to the indigenous products by dumping operations. Now is the best time to revive all these vanquished industries. Attempts are now being made to recover the losses of territories as in Cachar under duress.

Unity Of Hill And Valley

All these times when Manipur was in the height of her power, Hill and Valley were one; and this one-ness defended Manipur against all invasions and thus, she could maintain her independence upto 1891, when the rest of India had already been conquered by the British. British conquest of this land resulted from the slackened unity between 'the Valley and Hills. Now, since the withdrawal of British paramountcy after 56 years' rule, we have reinforced the union of Hill and Valley; and best opportunities have reached us in order to work hand-in-hand towards achieving the common object of progress and national prosperity. The relation has become one of complete understanding and active sympathy and it must be so for Manipur is one and the hopes and aspirations of Hill and Valley are identical.

Manipur's Position In The Dominion Of India

These are great days; great destinies are gather-ing momentum; great ideas are occupying the hearts of men and great causes have aroused great enthusiasms and great sacrifices. Amid the interplay of world forces India comes out stronger than before and Manipur too as a component part is the same with the removal of artificial hedge between Hill and Valley. I believe the sons of Manipur will be alive to it. In the large galaxy of heroes in the imperishable roll of honour, there were and there are now and there will never cease to be, beloved Manipuri names testifying to the fact that our people would rather die unsullied than outlive the disgrace of surrender to any measure that work prejudicial to the preservation of the separate entity of the State, while fostering the good and cordial relations with the Dominion of India I am confident that the members of the Assembly will please see that this fair record is never breached.

State Finance based on a Sound Economic Planning

A strong government means a government with a strong financial infrastructure. In order to ensure finance a well thought out economic planning is necessary. I lay special stress on it, for a totally unplanned economy is a totally unplanned nation. I invite your attention to the inevitability of this economic planning so that you may be prepared to cope with it before the State finance assumes threatening proportions. The inexhaustible sources of her economic potentialities are already there, as it is provided by nature. What is only wanting is the modus operandi for increased productivity. The people will have to cultivate thriftiness, since the present rate of the flowing out of people's money in exchange of less important articles will jeopardize, at no distant date, the economic status of the individuals constituting the nation.

If the entire area of the State is properly explored and the sources are tapped, there is every possibility of the State's one crore of rupees as annual income enabling us at the same time to produce all our requirements. Places like Jiribam and Moreh, if properly industrialised with the running of saw-mills, cotton mills, paper mills, etc., will bring huge incomes to the State. Manipur has been declared the best place for installing hydro-electric power on commercial basis too. There should be industrial centres at strategic places–Hill and Valley–equipped with modern techniques and modern technology. I shall be glad to see Hill towns properly electrified, springing up. I desire that prior consideration be given to the construction of roads in the hills in order to facilitate communications. There are now only one lakh of 'Paris' of land thrown open to cultivation. There is ten times the space still available for agricultural operations. Plantation of betal-nut, tea and jute also may be introduced successfully in the now neglected but fertile soils. Export should outpace the import. During my long tours in the hills, I was deeply impressed with the yet surviving cottage industries which can stand any competition. The consequences of the war heavily told upon the raw materials. Charkha and handlooms have been the essential marriage gifts to our mo hers and sisters. In re-dressing economic problems these should be developed. As regards the raw materials, eastern Kasom still stands a living testimony. Khaddar is not a new thing here, but rather–I should say–it originated from Manipur. The silk industry from cocoons is a long standing industry here. The Muga of Khurkhul remains unrivalled and unapproached in India. Rural reconstruction schemes be implemented in order to achieve all-round progress.

Education

Prior, consideration should also be given to education. I wish to see at lea.st a High School in each of the Tahsils and important' centres in the hills with good communications. While pursuing our future educational policy, we should be prepared to select more practical courses to suit the mode of our national life. Indigenous science of medicine and treatment should also be promoted. Fine Arts also has received the appreciation of the world. It attracted the world-poet Rabindranath Tagore also, Sir Akbar Hydari, His Excellency, the Governor of Assam, also highly encouraged it.

Our Part in the Individual and Collective Responsibility for Frontier Protection and Defence of Manipur in Particular

Manipur should be made militarily strong as, in the contemporary world, militia is the noblest estate of mankind. In the history and accounts of 1874, written by Col. Johnstone, you will come across that Manipuri nation constituted the best armies in the world. In January 1947, I apprised India Government through the Chamber of Princes, of the need for a strong Eastern frontier protecting military installations in order to safeguard Manipur in particular and India as a whole, as a result of the withdrawal of British paramountcy from India and Burma. But India Government have not yet communicated their final plan in this matter.

Now as from today, you have joined responsible government. I feel it my duty to point out, in the interest of this country and Commonwealth that it should be our individual and collective responsibility to work out a scheme for frontier protection, in collaboration with the Central Government. The present situation calls for immediate action in this respect. Since we are sincere, I am confident that the Central Government will give us proper and timely assistance with subsidies and finance and equipment. In the meantime with what little finance, we have, we should raise our small force to an efficient standard as homeguard without further delay. The inherent warlike spirit of the Manipuris should be maintained for the noble cause of India. Peace here in Manipur will be a fine contribution to India Government, while they are moulding India's destiny to enable her to stand up as one of the most powerful nations in the world.

The paramount importance of the strategic location of Manipur for defence and security reasons of India deserves the careful attention of enlightened governments; and we should be mindful of the benefits that we may render to our Dominion from her position, if we seriously depend on it.

Time is now roughly going on in the world and as such, it is a critical phase of Manipur too. Nevertheless, it is the best time for you to serve your mother, country while she needs and to prove that you are her worthy sons. You are to show to the world and to your people, the achievements or progress of the welfare of the people under this new government with clear results of a change for the better. Otherwise, the name "Responsible government" may not be worth the name and the next step will be for the Constitutional Ruler to judge.

Our Guiding Principles

We should not forget that we are Manipuris, destined to live in

Manipur and die in Manipur. Providence has bound our fates together indissolubly.

Leaders of the nation! While undertaking to finance economic planning–long and short–for the material progress of the country in the nearest future, I want to make it a watchword that the means of increasing the State revenues from other sources are devised without causing hardship or burden to the people. I expect that everyone of you, as a part of his obligation, will take care to acquaint himself with the mode of life on hills too by studying the four walls of the territory of Manipur, he is now going to govern, Send students abroad to foreign lands that they may bring back knowledge with which they may do service to her, but they should not bring ideas and ways of life that do not always conform to the environment of Manipur.

I wish to see you all becoming great and noble not for your own sake but noble for the country–to make Manipur great, a greater Manipur to enable her to stand up with head erect among the nations, as she did in the past. Work that Manipur may prosper. Suffer that Manipur may rejoice. Unite that Manipur may be strong.

Do the right that you may be brave. Whether you use pen or tongue, let your hearts be pure; and your resolve, manly. The prosperity of a nation does not depend on the number of buildings, but it depends on the number of men of character–that is, radiant man of well-disciplined mould. I shall always remain aglow with the splendour of destiny of our people.

In fine, I call upon you all to take this as your motto and to join with me in a fervent prayer for the well-being of our Motherland.

APPENDIX VI

The Manipur Merger Agreement

Agreement made this twentyfirst day of September 1949 between the Governor General of India and His Highness the Maharaja of Manipur.

Whereas in the best interests of the State of Manipur as well as of the Dominion of India it is desirable to provide for the administration of the said State by or under the authority of the Dominion Government:

It is Hereby Agreed as Follows:

Article I

His Highness the Maharaja of Manipur hereby cedes to the Dominion Government full and exclusive authority, jurisdiction and powers for and in relation to the governance of the State and agrees to transfer' the administration of the State to the Dominion Government on the fifteenth day of October 1949 (hereinafter referred to as "the said day").

As from the said day the Dominion Government will be competent to exercise the said powers, authority and jurisdiction in such manner and through such agency as it may think fit.

Article II

His Highness the Maharaja shall continue to enjoy the same personal rights, privileges, dignities, titles, authority over religious observances, customs, usages, rites and ceremonies and institutions charge of the same in the State, which he would have enjoyed had this agreement not been made.

Article III

His Highness the Maharaja shall with effect from the said day be entitled to receive for his life-time from the revenues of the State annually for his Privy Purse the sum of Rupees three lakhs, free of all taxes.

This amount is intend d to cover all the expenses oft e Ruler and his family, including expenses on account of his personal staff and armed guards, maintenance of his residences, marriages and other ceremonies etc. and the allowances to the Ruler's relations who on the date of execution of this agreement were in receipt of such allowances from the revenues of the State, and will neither be increased nor reduced for any reason whatsoever.

The Government of India undertake that the said sum of Rupees three lakhs shall be paid to His Highness the Maharaja in four equal installments in advance at the beginning of each quarter from the State treasury or at such other treasury as may be specified by the Government of India.

Article IV

His Highness the Maharaja shall be entitled to the full ownership, use and enjoyment 'of all private properties (as distinct from State properties) belonging to him on the date of this agreement.

His Highness the Maharaja will furnish to the Dominion Government before the first January 1950 an inventory of all the immovable property, securities and cash balance held by him as such private property.

If any dispute arises as to whether any item of property is the private property of His Highness the Maharaja or State property, it shall be referred to a judicial officer qualified to be appointed as a High Court Judge, and the decision of that officer shall be final and binding on both parties:

Provided that His Highness the Maharaja's right to the use of the residences known as "Red-lands" and "Les Chatalettes" in Shillong, and the property in the town of Gauhati known as "Manipuri Basti" shall not be questioned.

Article V

All the members of His Highness's family shall be entitled to an the personal rights, privileges, dignities and titles enjoyed by them whether within or outside the territories of the State, immediately before the, 15th August, 1947.

Article VI

The Dominion Government guarantees the succession, according to law and custom, to the gaddi of the State and to His Highness, the Maharaja's personal rights, privileges, dignities, titles, authority over religious observances, customs usages, rites and ceremonies and institutions incharge of the same in the State.

Article VII

No enquiry shall be made by or under the authority of the Government of India, and no proceedings shall lie in any Court in Manipur, against His Highness the Maharaja whether in a personal capacity or otherwise in respect of anything done or omitted to be done by him or under his authority during the period of his administration of that State.

Article VIII

(1) The Government of India hereby guarantees either the continuance in service of the permanent members of the Public Services of Manipur on conditions which will be not less advantageous than those on which they were serving before the date on which the administration of Manipur is made over to the Government of India or the payment of reasonable compensation.

(2) The Government of India further guarantees the continuance of pensions and leave salaries sanctioned by His Highness the Maharaja to servants of the State who have retired or proceeded on leave preparatory to retirement, before the date on which the Administration of Manipur is made over to the Government of India.

(3) The Government of India also undertake to make suitable provisions for the employment of Manipuris in the various branches of Public Services, and in every way encourage Manipuris to join them. They also undertake to preserve various laws, customs and conventions prevailing in the State pertaining to the social, economic and religious life of the people.

Article IX

Except with the previous sanction of the Government of India no proceedings, civil or criminal, shall be instituted against any' person in respect of any act done or purporting to be done in the execution of his duties as a servant of the State before the day on which the administration is made over to the Government of India.

In confirmation whereof Mr. Vapal Pangunni Menon, Adviser to the Government of India in the Ministry of States, has appended his signature on behalf and with the authority of the Governor General of India and His Highness Maharaja Bodhchandra Singh, Maharaja of Manipur has appended his signature on behalf of himself, his heirs and successors.

September 21, 1949.

Bodhchandra Singh,
Maharaja of Manipur.

V.P. Menon,
Adviser to the Government of India, Ministry of State.

Sri Prakasa
Governor of Assam, Shilong

Ammunitions recovered

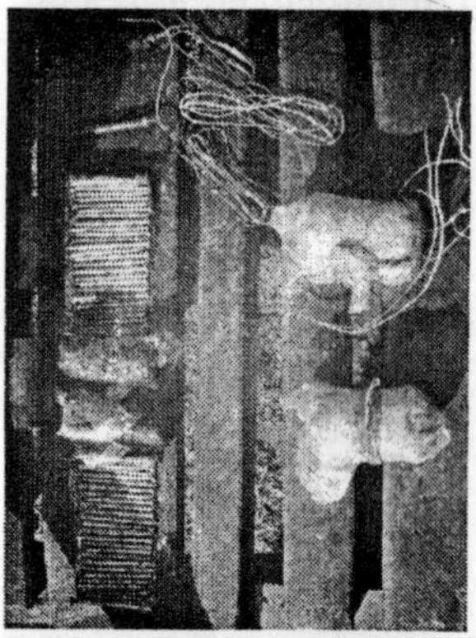

Another set of IED

Boat Operation in Loktak

Boat Operation

Combined Operation in valley Saji Tampak

Police & CO of a Battalion taking account of captured weaons

Recovery of IED materials

Electronic IED's

Fighting Bunker of UG Outfits.

IED recover deep in jungle

Living room and Kitchen, Loktak Lake

New samtal blast (IED blown off leg

Operation at Loktak

Putting up OP in thick of phumdies

Troops providing Medical evlauation

Recovery of electronics devices

Search Operation of a Hotel

Troops finally reach Old Samtal IB with Burma

Troops in search of death body of insurgents in Parbung

Index

■■■

Article VI

The Dominion Government guarantees the succession, according to law and custom, to the gaddi of the State and to His Highness, the Maharaja's personal rights, privileges, dignities, titles, authority over religious observances, customs, usages, rites and ceremonies and institutions in charge of the same in the State.

Article VII

No enquiry shall be made by or under the authority of the Government of India, and no proceedings shall lie in any Court in Manipur, against His Highness the Maharaja whether in a personal capacity or otherwise in respect of anything done or omitted to be done by him or under his authority during the period of his administration of that State.

Article VIII

(1) The Government of India hereby guarantees either the continuance in service of the permanent members of the Public Services of Manipur on conditions which will be not less advantageous than those on which they were serving before the date on which the administration of Manipur is made over to the Government of India or the payment of reasonable compensation.

(2) The Government of India further guarantees the continuance of pensions and leave salaries sanctioned by His Highness the Maharaja to servants of the State who have retired or proceeded on leave preparatory to retirement, before the date on which the administration of Manipur is made over to the Government of India.

(3) The Government of India also undertake to make suitable provisions for the employment of Manipuris in the various branches of Public Services, and in every way encourage Manipuris to join them. They also undertake to preserve various laws, customs and conventions prevailing in the State pertaining to the social, economic and religious life of the people.